Internet of Things
Global Technological
and Societal Trends

RIVER PUBLISHERS SERIES IN COMMUNICATIONS

Consulting Series Editors

MARINA RUGGIERI
University of Roma "Tor Vergata"
Italy

HOMAYOUN NIKOOKAR
Delft University of Technology
The Netherlands

This series focuses on communications science and technology. This includes the theory and use of systems involving all terminals, computers, and information processors; wired and wireless networks; and network layouts, pro-contentsols, architectures, and implementations.

Furthermore, developments toward new market demands in systems, products, and technologies such as personal communications services, multimedia systems, enterprise networks, and optical communications systems.

- Wireless Communications
- Networks
- Security
- Antennas & Propagation
- Microwaves
- Software Defined Radio

For a list of other books in this series, see final page.

Internet of Things Global Technological and Societal Trends

Dr. Ovidiu Vermesan
SINTEF, Norway

Dr. Peter Friess
EU, Belgium

Aalborg

Published, sold and distributed by:
River Publishers
PO box 1657
Algade 42
9000 Aalborg
Denmark
Tel.: +4536953197

www.riverpublishers.com

ISBN: 978-87-92329-73-8
© 2011 River Publishers

All rights reserved. No part of this publication may be reproduced, stored in a retrieval system, or transmitted in any form or by any means, mechanical, photocopying, recording or otherwise, without prior written permission of the publishers.

Dedication

"It is an exciting time where the only limits you have are the size of your ideas and the degree of your dedication"

Acknowledgement

The editors would like to thank the European Commission for their support in the planning and preparation of this book. The recommendations and opinions expressed in the book are those of the editors and contributors, and do not necessarily represent those of the European Commission.

Ovidiu Vermesan
Peter Friess

"This cluster book on Internet of Things is another important step on the long European path to a more competitive, dynamic, inclusive and user-friendly knowledge-based economy. I commend the distinguished contributors from all over the world who have made production of this book a major priority, and I am grateful for their leadership and skill. At the same time, I want to acknowledge the positive role played by the European Commission and more particularly, all those in DG Information Society and Media who, since 2006, have devoted their time, energy and ideas to the further development of the Internet of Things vision. This cluster book demonstrates that there is too much at stake with the future of the Internet of Things to not build the durable cooperation among all key players, which is a prerequisite for ensuring the best decisions possible before moving forward on technology, applications, use cases and implementations."

Gérald Santucci
Head of Unit, European Commission

Editors Biography

Dr. Ovidiu Vermesan holds a Ph.D. degree in microelectronics and a Master of International Business (MIB) degree. He is Chief Scientist at SINTEF Information and Communication Technology, Oslo, Norway. His research interests are in the area of analog and mixed-signal ASIC Design (CMOS/BiCMOS/SOI) with applications in measurement, instrumentation, high-temperature applications, medical electronics and integrated sensors; low power/low voltage ASIC design; and computer-based electronic analysis and simulation. Dr. Vermesan received SINTEFs 2003 award for research excellence for his work on the implementation of a biometric sensor system. He is currently working with projects addressing nanoelectronics integrated systems, communication and embedded systems, integrated sensors, wireless identifiable systems and RFID for future Internet of Things architectures with applications in green automotive, internet of energy, healthcare, oil and gas and energy efficiency in buildings. He is actively involved in the activities of the European Technology Platforms ENIAC (European Nanoelectronics Initiative Advisory Council), ARTEMIS (Advanced Research & Technology for EMbedded Intelligence and Systems), EPoSS (European Technology Platform on Smart Systems Integration). Dr. Vermesan is the coordinator of the IoT European Research Cluster (IERC) of the European Commission, actively participated in EU FP7 Projects related to Internet of Things.

Dr. Peter Friess holds a Ph.D. in engineering with emphasis on business reengineering, project management and self-organising systems, and a diploma degree in aerospace and aeronautics engineering from the Technical University of Munich. Peter Friess has written several articles and participated in the improvement of world-wide project management approaches. He is responsible for the management of the European Internet of Things research cluster with emphasis on identification and sensing technologies, connected objects, architectural frameworks and new application scenarios like Smart Cities. He is also responsible for the management of the first Pan-European Thematic Network for Radio Frequency Identification Technologies and the international co-operation on the Internet of Things, in particular with Asian countries. Prior to his present activities, he was e-business senior consultant at IBM, consulting major automotive and utilities companies in Europe, and before that he was active as IT-manager at Philips Semiconductors. He started his career as researcher for business reengineering at the Bremen Institute for Applied Work Sciences.

Foreword

We are happy to present to you this second edition of the IERC — the Internet of Things European Research Cluster. Many developments have occurred in the past year. The Internet of Things has gained worldwide recognition as a key concept both for addressing the huge societal challenges we face and opening up tremendous business opportunities in applications areas such as Smart Cities, Assisted Living, eHealth, Intelligent Manufacturing and Intelligent Transport and Logistics.

Researchers and service providers are concentrating on the important horizontal dimension of the Internet of Things — as the objects move from one application sector to another, the mechanisms for sensing and actuation, connection, communication, security and privacy should function seamlessly and without redundancy. An additional research priority is lowering of overall energy consumption. Finally, work is on-going on aspects of decentralised intelligence, self-organising systems, new interfaces and the co-existence of real and virtual worlds.

The Cluster has also extended its outreach activities to the political and international level. After two waves of newly selected FP7 EU IoT research projects and improved co-ordination with various national initiatives, the research base is sound and moving in the right direction. The Cluster is assuming its responsibilities in contributing to wider public policy and innovation issues by having a permanent seat in the Commission's IoT Expert Group, backing the work of bi-lateral international expert groups and national advisory boards while linking to European Technology Platforms and providing position papers inputs in the context of preparations for the 8th Framework Programme on European Research.

More than ever the Internet of Things is a subject of global interest and when preparing the concept for this Clusterbook, we decided to invite our international partners to share with all of us their points of view. As the world becomes flatter and hyper-connected, we are conscious of the need to maintain Europe's strong position — less in a sense of competition but more in a common understanding.

If you would allow us a wish, we would encourage all our Internet of Things researchers to advance this year beyond state-of-the-art, all entrepreneurs to make the Internet of Things a commercial success, and all public and private decision makers to establish the appropriate innovation landscape.

Special thanks are expressed to Ovidiu Vermesan, his entire team and all contributors for making this Clusterbook a new milestone in thinking about Internet of Things.

Brussels, March 2011
Peter Friess

Contents

Foreword   xi

1 European Research Cluster on the Internet of Things   1
Ovidiu Vermesan and Peter Friess

1.1	Introduction	1
1.2	IoT European Research Cluster	2
1.3	IoT Activity Chains	3
1.4	IoT European Research Cluster Role	4
1.5	Conclusion	6
	References	7

2 Internet of Things Strategic Research Roadmap   9
Dr. Ovidiu Vermesan, Dr. Peter Friess, Patrick Guillemin, Sergio Gusmeroli, Harald Sundmaeker, Dr. Alessandro Bassi, Ignacio Soler Jubert, Dr. Margaretha Mazura, Dr. Mark Harrison, Dr. Markus Eisenhauer, Dr. Pat Doody

2.1	Internet of Things Conceptual Framework	10
2.2	Internet of Things Vision	10
2.3	Technological Trends	18
2.4	IoT Applications	20
2.5	Technology Enablers	22
2.6	Internet of Things Research Agenda, Timelines and Priorities	26
2.7	Future Technological Developments	44
2.8	Internet of Things Research Needs	46
	References	51

3	**The Internet of Things: The Way Ahead**	**53**
	Dr. Gérald Santucci	
3.1	Little Stones on the Road…	53
3.2	The State of the Art of Internet of Things Research in Europe	54
3.3	Towards an EU Policy Framework	74
3.4	Outlook	96
	References	98

4	**Between Big Brother, Matrix and Wall Street: The Challenges of a Sustainable Roadmap for the Internet of Things**	**101**
	Dr. Alessandro Bassi	
4.1	Introduction	101
4.2	Sociological Issues	102
4.3	Technological Challenges	103
4.4	Business Considerations	106
4.5	Future Works	107

5	**2011: A Decisive Year**	**109**
	Rob van Kranenburg	
5.1	Introduction	109
5.2	Impact	109
	References	114

6	**Internet of Things — From Ubiquitous Computing to Ubiquitous Intelligence Applications**	**115**
	Prof. Ken Sakamura and Chiaki Ishikawa	
6.1	Open Code Architecture: Ubiquitous ID Architecture	115
6.2	Ubiquitous Computing	115
6.3	Ubiquitous ID Architecture	117
6.4	Ucode	120
6.5	Features of Ucode	120
6.6	Ucode Tags	121

6.7	Ubiquitous Communicator (UC)	122
6.8	Ucode Resolution Server and Information Server	123
6.9	UID 2.0 — Realization of Richter Ubiquitous Computing World Based on ucR	124
6.10	Ucode Usage Examples	124
6.11	Application for Places	125
6.12	Tokyo Ubiquitous Technology Project	125
6.13	Tokyo Ubiquitous Technology Project in Ginza	126
6.14	Governance and Sustainability: Participation by Private Companies	126
6.15	Metropolitan Government Ubiquitous Sightseeing Guide	126
6.16	"Portable Information System," Ueno Zoological Gardens	127
6.17	Hama-rikyu Gardens Ubiquitous Guidance Service	128
6.18	*kokosil*, Location Information Portal	128
6.19	Use of Ucode Location Information Systems Spread to Many Regions	129
6.20	Application of Ucodes for Objects	134
6.21	Ucode Application in the Management of Historical Information of Houses	135
6.22	Traceability Management System of Housing Components	136
6.23	Cyber Concrete	137
6.24	Summary	139
	References	140

7 Technologies, Applications, and Governance in the Internet of Things — **143**

Prof. Lirong Zheng, Hui Zhang, Weili Han, Xiaolin Zhou, Jing He, Zhi Zhang, Yun Gu, and Junyu Wang

7.1	Overview	143
7.2	Key Technologies	144
7.3	Technical Challenges of the Internet of Things	174
7.4	Conclusion	175
	References	176

8 Mobile Devices Enable IoT Evolution From Industrial Applications to Mass Consumer Applications — 179

Prof. Jian Ma and Long Cheng

8.1	Introduction	179
8.2	IoT Mobile Device	180
8.3	IoT Applications in Current Mass Consumer Market	181
8.4	Future IoT Applications in Mass Consumer Market	186
8.5	Conclusion	192
	References	192

9 Opportunities, Challenges for Internet of Things Technologies — 195

Dr. José Roberto de Almeida Amazonas

9.1	Introduction	195
9.2	Architecture Models	201
9.3	Network Technology	215
9.4	Discovery and Search Engines	222
9.5	Security and Privacy	226
9.6	Application Areas and Industrial Deployment	232
9.7	Governance and Socio-economic Ecosystems	233
9.8	Contribution to an Iot Global Vision: Opportunities and Challenges	235
	References	239

10 Virtualization of Network Resources and Physical Devices in Internet of Things Applications — 241

Dr. Sangjin Jeong, Dr. Myung-Ki Shin and Dr. Hyoung-Jun Kim

10.1	Introduction	241
10.2	Overview of Network Virtualization	242
10.3	Challengens for Network Virtualization	246
10.4	Requirements of Network Virtualization	248
10.5	Applicability of Network Virtualisation	252
10.6	Conclusions	255
	References	255

11 Interoperability, Standardisation and Governance in the Era of Internet of Things (IoT)   257

Vandana Rohokale, Rajeev Prasad, Dr. Neeli Prasad and Prof. Ramjee Prasad

11.1	Introduction	257
11.2	Interoperability in the IoT	258
11.3	Standardisation For IoT	262
11.4	Specific Issues in IoT Standardisation	265
11.5	Security and Privacy Issues	266
11.6	Architecture and Governance of the IoT	270
11.7	Middleware or Software Platform for IoT	273
11.8	Network Technology	275
11.9	Application Areas and Industrial Deployment	277
11.10	Socio-economic Ecosystems	281
11.11	Conclusions and Future Work	282
	References	283

12 IoT Validation and Interoperability IoT — IPv6 and M2M   287

Marylin Arndt, Philippe Cousin, Patrick Grossetete, Latif Ladid, and Sébastien Ziegler

12.1	M2M: A Set of Enabling Technologies Paving the Way From the Internet of Services Towards the Internet of Things	287
12.2	The "Internet of Things" Based on IPv6	297
12.3	Paving the Way to Smart IPv6 Buildings	304
12.4	Validation and Interoperability Challenges for IoT	308
12.5	Conclusions	311
	References	312

Index   315

1
European Research Cluster on the Internet of Things

Dr. Ovidiu Vermesan[1] and Dr. Peter Friess[2]

[1] *SINTEF, Norway*
[2] *European Commission, Belgium*

1.1 Introduction

The Internet of Things (IoT) is formed by the networked interconnection of everyday objects. It is involving self configuring wireless networks of sensors that create a world where everything in it sends information to other objects and to people. This world, in which everything is tagged and communicating, provides information and knowledge that enable us to live better lives and to easier solve problems.

Internet of Things enables the objects in our environment to be active participants, i.e., they share information with other members of the network or with any other stakeholder and they are capable of recognizing events and changes in their surroundings and of acting and reacting autonomously in an appropriate way.

In this context the research and development challenges to create a smart world where the real, digital and the virtual worlds are converging to create smart environments that would make energy, transport, cities and many other areas more intelligent are enormous.

The development of certain enabling technologies such as nanoelectronics, communications, sensors, smart phones, embedded systems, cloud computing and software will be essential to support important future IoT product innovations affecting many different industrial sectors.

Today many European projects and initiatives address the Internet of Things technologies and knowledge. Given the fact that the topics can be

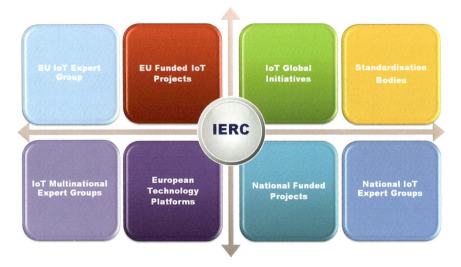

Fig. 1.1 IERC coordinates the convergence of ongoing activities [2].

highly divers and specialized, it exists a strong need for integration of the individual results. Knowledge integration, in this context is conceptualized as the process through which disparate, specialized knowledge located in multiple projects across Europe is combined, applied and assimilated.

1.2 IoT European Research Cluster

In this new dynamic environment the European Commission created the IoT European Research Cluster — European Research Cluster on the Internet of Things (IERC) [1]. The Cluster's role is pivotal due to its goal to allow projects to be more productive and innovative than they could be in isolation. And the IERC is important because it reduces the barriers to entry for new value creation and exploitation, dissemination and deployment of ideas and research results. The aim is to address the large potential for IoT-based capabilities in Europe and to coordinate the convergence of ongoing activities [2], as can be seen in Fig. 1.1.

The Cluster plays this role, because the projects, organizations and companies participating in the IoT Cluster share the following characteristics:

- Common Activity Chains; they have close cooperation around a number of common activities that allows positive spillovers and sharing of common resources.

- Linkages; the IERC activities share a common goal, for example, a common vision for IoT, common research strategic agenda, final market demand, etc.
- Interactions; there is a close working relationship on related issues of interest for several projects, and high level of active interaction through common seminars and workshops.
- Critical mass; there are sufficient number of participants (around 30 projects) present for the interactions which has a relevant impact on the performance of the actors involved.

1.3 IoT Activity Chains

Managing the knowledge integration process between the projects in the Cluster is a crucial task. The challenge arises not only because knowledge is often dispersed, differentiated and embedded in various projects but also because the projects and the partners have their own agendas within organizations that are intrinsically different, that may possess diverse competencies and conflicting interests [3].

In this context the activity chains (ACs) are created to favour close cooperation between the IoT Cluster projects and to form an arena for exchange of ideas and open dialog on important research challenges. The activity chains are defined as work streams that group together partners or specific participants from partners around well defined technical activities that will result into at least one output or delivery that will be used in addressing the IERC objectives.

The Cluster coordinator, in cooperation with the European Commission coordinator, has defined 15 activities chains that are grouped in the following 5 main areas:

1. Architecture approaches, models, naming, search, discovery
2. Governance issues, privacy and security
3. Links to national, European and international initiatives
4. Interoperability, standardisation, dissemination, exploitation
5. Coordination of the Strategic Research Agenda (SRA) at the European level in the global view.

For each activity chain, a project from the new projects started in 2010 was assigned as activity chain coordinator, and the project coordinator has nominated a person from the project as activity chain responsible. The activities to support the IERC are described and included in the description of work (DoW) of the respective projects. The activity chain responsible persons report to the IERC coordinator and their role is to:

- Organize the work according to the schedule in the respective activity chain.
- Coordinate the participation and the contribution of the different projects in addressing and fulfilling the tasks assigned to the activity chain.
- Supervise and facilitate the information exchange and coordinate the action plan, deliverables and outputs to the Cluster.
- The responsible person for the activity chain will be present at all the cluster meetings. In special cases he/she will delegate another person to participate to the meetings.
- The Activity Chain responsible will present the activities at the IERC meetings. The Cluster will plan the presentation of 2–4 ACs at one meeting.

The projects and partners/persons that are interested to join individual activity chains contact the activity chain responsible and agree how to participate, contribute and provide support to the respective activity chain.

1.4 IoT European Research Cluster Role

The IoT European Research Cluster (IERC) aims at defining the IoT Research, Development and Innovation (R+D+I) challenges at the European level in view of global developments. The IERC brings together European projects in the field of IoT technology funded by the FP7 ICT programme, thus underpinning the basic multidisciplinary science. The rationale for the IERC is to address the large potential for IoT-based capabilities in Europe — coordinate/encourage the convergence of ongoing work addressing the most important challenges — to build a broadly based consensus on the ways to implementation and adoption of IoT in Europe. The cluster yielded results that led to the publishing of the Internet of Things Strategic Research Roadmap [4] and the Cluster Book "Vision and Challenges for Realising the Internet of Things" [5]. These activities require in-depth involvement of each project and an active coordination of the activities among the various projects involved in research and development in the area of IoT and Future Internet, with a strong international perspective and an application oriented goal.

The Cluster links its activities with the activities of the IoT Expert Group to minimise overlaps and maximize synergies in order to contribute to the overall objectives of an IoT strategy for Europe that include:

- To align the activities on focused and shared research, structured knowledge and networking of partners in the area of IoT R+D+I;
- To establish a cooperation platform that develops a research vision for IoT activities in Europe and that becomes a major entry and contact point for IoT research in the world;
- To reinforce international cooperation with emerging economies in order to find common solutions for common challenges;
- To define an international strategy for cooperation in the area of IoT research based on an overview of research priorities at global level;
- To strengthen the dialogue between cluster projects and initiatives outside the cluster in order to identify cooperation opportunities and to foster cross-fertilization and impact creation;
- To contribute to the Digital Agenda's goals in encouraging cross-stakeholder knowledge exchange around a common research agenda and in identifying industry-led innovative application areas for IoT;
- To investigate how to enhance IERC's visibility in order to become a globally recognised voice in the area of IoT;
- To organise debates/workshops/conferences leading to awareness raising and a better understanding of IoT and Future Internet Science and Technology, in particular vis-à-vis socio-economic challenges;
- To prepare the annual Cluster work programme including objectives, goals, achievements and indicators, and coordinating the work with the projects participating in the Cluster accordingly;
- To coordinate and organize bi-annual Cluster meetings aligned with important IoT events/conferences/workshops;

- To coordinate and align the Strategic Research Agenda document at European level with the developments at global level and publish annually the Cluster Book.

The long term ambition of the IoT European Research Cluster is to strengthen the competitiveness of Europe in this area through:

- generating new applications and businesses based on IoT technology;
- facilitating the use of research and development results by the network partners and market players in general;
- fostering cooperation of all IoT stakeholders at European and international level.

The envisioned results obtained after 5 years will be at two levels: at fundamental level through the strategic research roadmap and identification of new applications for IoT technology, and at implementation level that defines business cases and business processes based on the IoT technology and that may give solutions to socio-economic challenges.

1.5 Conclusion

The IoT Cluster development is focusing on how to activate the existing base of European and national projects, companies and institutions to jointly participate in common activities in much more effective ways. The IERC has become during the last years conceptually more consistent and has worked to harmonise the specific policy priorities while connecting with the national programs. European and national authorities are enablers in spreading the IoT concepts, with the IoT Cluster seen as part of a wider competitiveness agenda. Efforts for Cluster and for national projects need to be better integrated. Focusing exclusively on a few projects limits the impact on overall European impact. Focusing on both European and national funded projects gives the opportunity to address a broader perspective, spreads the impact widely to make a material difference to groups of projects and companies that share Cluster specific objectives to further growth and innovation. Successful competitiveness efforts aim to achieve a European leading role in the area of IoT through a combination of Cluster specific and crosscutting initiatives.

References

[1] IERC — European Research Cluster on the Internet of Things, http://www.internet-of-things-research.eu/.

[2] IERC — European Research Cluster on the Internet of Things, "Coordinating and building a broadly based consensus on the ways to realise the Internet of Things (IoT) vision in Europe," Poster, http://www.internet-of-things-research.eu/pdf/Poster_IERC_A0_V01.pdf.

[3] M. Bhandar, "Knowledge Clusters: Dealing with a Multilevel Phenomenon," *U21Global-Working Paper*, No. 007/2008, http://www.u21global.edu.sg/PartnerAdmin/ViewContent?module=DOCUMENTLIBRARY&oid=157453.

[4] SRA, "Internet of Things Strategic Research Roadmap," September 2009, http://www.internet-of-things-research.eu/pdf/IoT_Cluster_Strategic_Research_Agenda_2009.pdf.

[5] CERP-IoT — Cluster of European Research Projects on the Internet of Things, "Vision and Challenges for Realising the Internet of Things," Mars 2010, http://www.internet-of-things-research.eu/pdf/IoT_Clusterbook_March_2010.pdf.

2

Internet of Things Strategic Research Roadmap

Dr. Ovidiu Vermesan[1], Dr. Peter Friess[2], Patrick Guillemin[3],
Sergio Gusmeroli[4], Harald Sundmaeker[5], Dr. Alessandro Bassi[6],
Ignacio Soler Jubert[7], Dr. Margaretha Mazura[8], Dr. Mark Harrison[9],
Dr. Markus Eisenhauer[10], Dr. Pat Doody[11]

[1] *SINTEF, Norway*
[2] *European Commission, Belgium*
[3] *ETSI, France*
[4] *TXT e-solutions, Italy*
[5] *ATB GmbH, Germany*
[6] *IoT-A Project, France*
[7] *ATOS Origin, Spain*
[8] *EMF, UK*
[9] *Institute of Manufacturing, University of Cambridge, UK*
[10] *Fraunhofer FIT, Germany*
[11] *Centre for Innovation in Distributed Systems, Institute of Technology, Ireland*

> "What most people need to learn in life is how to love people and use things instead of using people and loving things."

> "It is not because things are difficult that we do not dare, it is because we do not dare that things are difficult."
>
> ***Seneca***

> "All things appear and disappear because of the concurrence of causes and conditions. Nothing ever exists entirely alone; everything is in relation to everything else."
>
> ***Hindu Prince Gautama Siddharta***

2.1 Internet of Things Conceptual Framework

Internet of Things (IoT) is an integrated part of Future Internet including existing and evolving Internet and network developments and could be conceptually defined as a dynamic global network infrastructure with self configuring capabilities based on standard and interoperable communication protocols where physical and virtual "things" have identities, physical attributes, and virtual personalities, use intelligent interfaces, and are seamlessly integrated into the information network.

In the IoT, "smart things/objects" are expected to become active participants in business, information and social processes where they are enabled to interact and communicate among them-selves and with the environment by exchanging data and information "sensed" about the environment, while reacting autonomously to the "real/physical world" events and influencing it by running processes that trigger actions and create services with or without direct human intervention.

Services will be able to interact with these "smart things/objects" using standard interfaces that will provide the necessary link via the Internet, to query and change their state and retrieve any information associated with them, taking into account security and privacy issues [1].

The IERC definition aims to coin the IoT paradigm and concept by unifying the different statements and many visions referred to as a "Things," "Internet," "Semantic," "Object Identification" oriented definitions of Internet of Things promoted by individuals and organisations around the world.

This enables a common vision for the deployment of independent federated services and applications, characterized by a high degree of autonomous data capture, event transfer, network connectivity and interoperability.

2.2 Internet of Things Vision

The vision of Future Internet based on standard communication protocols considers the merging of computer networks, Internet of Things (IoT), Internet of People (IoP), Internet of Energy (IoE), Internet of Media (IoM), and Internet of Services (IoS), into a common global IT platform of seamless networks and networked "smart things/objects".

IoE is defined as a dynamic network infrastructure that interconnects the energy network with the Internet allowing units of energy (locally generated, stored, and forwarded) to be dispatched when and where it is needed. The related information/data will follow the energy flows thus implementing the necessary information exchange together with the energy transfer.

IoS is denoting a software based component that will be delivered via different networks and Internet. Research on SOA, Web/enterprise 3.0/X.0, enterprise interoperability, service Web, grid services and semantic Web will address important bits of the IoS puzzle, while improving cooperation between service providers and consumers.

IoM will address the challenges in scalable video coding and 3D video processing, dynamically adapted to the network conditions that will give rise to innovative applications such as massive multiplayer mobile games, digital cinema and in virtual worlds placing new types of traffic demands on mobile network architectures.

IoP interconnects growing population of users while promoting their continuous empowerment, preserving their control over their online activities and sustaining free exchanges of ideas. The IoP also provides means to facilitate everyday life of people, communities, organizations, allowing at the same time the creation of any type of business and breaking the barriers between information producer and information consumer (emergence of prosumers).

IoT together with the other emerging Internet developments such as Internet of Energy, Media, People, Services, Business/Enterprises are the backbone of the digital economy, the digital society and the foundation for the future knowledge based economy and innovation society. IoT developments show that we will have 16 billion connected devices by the year 2020 [1], which will average out to six devices per person on earth and to many more per person in digital societies. Devices like smart phones and machine to machine or thing to thing communication will be the main drivers for further IoT development.

By 2015, wirelessly networked sensors in everything we own will form a new Web. But it will only be of value if the "terabyte torrent" of data it generates can be collected, analyzed and interpreted [6].

The first direct consequence of the IoT is the generation of huge quantities of data, where every physical or virtual object connected to the IoT may have a digital twin in the cloud, which could be generating regular updates. As a result, consumer IoT related messaging volumes could easily reach between 1.000 and 10.000 per person per day [2].

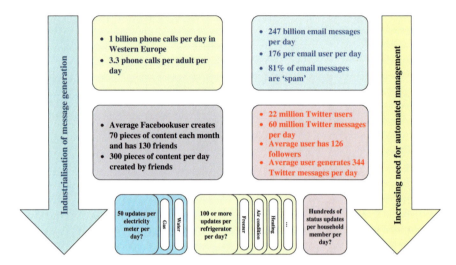

Fig. 2.1 The industrialisation of message generation [1].

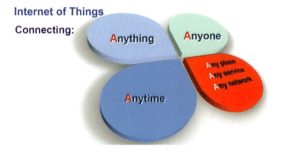

Fig. 2.2 Internet of things — 6A connectivity.

The IoT contribution is in the increased value of information created by the number of interconnections among things and the transformation of the processed information into knowledge for the benefit of mankind and society.

The Internet of Things could allow people and things to be connected Anytime, Anyplace, with Anything and Anyone, ideally using Any path/network and Any service. This is stated as well in the ITU vision of the IoT, according to which: "From anytime, anyplace connectivity for anyone, we will now have connectivity for anything" [4].

The vision of what exactly the Internet of Things will be, and what will be its final architecture, are still diverging.

2.2 Internet of Things Vision 13

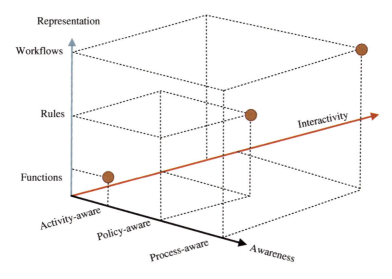

Fig. 2.3 Smart object dimensions: activity, policy and process aware [12].

A future network of networks could be laid out as public/private infrastructures and dynamically extended and improved by edge points created by the "things" connecting to one another. In fact, in the IoT communications could take place not only between things but also between people and their environment.

The vision of an Internet of Things built from smart things/objects needs to address issues related to system architecture, design and development, integrated management, business models and human involvement. This vision will have to take into account the integration of legacy systems and communications. Topics like the right balance for the distribution of functionality between smart things and the supporting infrastructure, modelling and representation of smart objects' intelligence, and programming models, are important elements that can be addressed by classifying smart object/things types as: *Activity-aware objects*, *policy-aware objects*, and *process-aware objects* [12]. These types represent specific combinations of three design dimensions with the aim to highlight the interdependence between design decisions and explore how smart objects can cooperate to form an "Internet of smart objects."

For instance, in [12] a vision of an IoT built by smart objects, able to sense, interprets, and react to external events is proposed. Within this vision, by capturing and interpreting user actions, smart items will be able to perceive and instruct their environment, to analyse their observations and to communicate with other objects and the Internet. This new Internet will co-exist and be intimately bound up with the Internet of information and services [13].

Utilizing real world knowledge on the networking levels, as well on service level will enable optimizing systems towards higher performance, better user experiences, as well as toward more energy efficiency.

Addressing elements such as Convergence, Content, Collections (Repositories), Computing, Communication, and Connectivity is likely to be instrumental in order to allow seamless interconnection between people and things and/or between things and things. The Internet of Things could imply a symbiotic interaction between the real/physical, world, and the digital/virtual world: physical entities have digital counterparts and virtual representation; things become context aware and they can sense, communicate, interact, exchange data, information and knowledge. 'Things' can only become context aware, sense, communicate, interact, exchange data, information and knowledge if they are suitably equipped with appropriate object-connected technologies; unless of course they are human 'things' or other entities with these intrinsic capabilities. In this vision, through the use of intelligent decision-making algorithms in software applications, appropriate rapid responses can be given to physical phenomena, based on the very latest information collected about physical entities and consideration of patterns in the historical data, either for the same entity or for similar entities. These create new opportunities to meet business requirements, create new services based on real time physical world data, gain insights into complex processes and relationships, handle incidents, address environmental degradation (for example pollution, disaster, tsunami, global warming), monitor human activities (health, movements, etc.), improve infrastructure integrity (energy, transport, etc.), and address energy efficiency issues (smart energy metering in buildings, efficient consumption by vehicles, etc.).

Everything from individuals, groups, communities, objects, products, data, services, processes could use the communication fabric provided by the smart things/objects. Connectivity will become in the IoT a kind of commodity, available to all at a very low cost and not owned by any private entity. In this context, there will be the need to create the right situation-aware development environment for stimulating the creation of services and proper intelligent middleware to understand and interpret the information, to ensure protection from fraud and malicious attack (that will inevitably grow as Internet becomes more and more used) and to guarantee privacy.

Capturing real world data, information and knowledge and events is becoming increasingly easier with sensor networks, social media sharing, location based services, and emerging IoT applications. The knowledge capturing and using is done in many cases at application level and the networks are mainly agnostic about what is happening around the terminals connected to the Internet.

Internet connectable consumer household devices will increase significantly in the next decade, with the computer network equipment that accounts for the majority of household devices, at about 75% in 2010, and declining to 25% by 2020 [1].

Embedding real world information into networks, services and applications is one of the aims of IoT technology by using enabling technologies like wireless sensor and actuator networks, IoT devices, ubiquitous device assemblies and RFID. These autonomous systems will "naturally" network with each other, with the environment, and the network infrastructure itself. New principles for self- properties, analysis of emerging behaviour, service platform approaches, new enabling technologies, as well as Web technology-based ideas will form the basis for this new "cognitive" behaviour.

Under this vision and making use of intelligence in the supporting network infrastructure, things will be able to autonomously manage their transportation, implement fully automated processes and thus optimise logistics; they have to be able to harvest the energy they need; they will configure themselves when exposed to a new environment, and show an "intelligent/cognitive" behaviour when faced with other things and deal seamlessly with unforeseen circumstances; and, finally, they might manage their own disassembly and recycling, helping to preserve the environment, at the end of their lifecycle.

The Internet of Things infrastructure allows combinations of smart objects (i.e., wireless sensors, mobile robots, etc.), sensor network technologies, and human beings, using different but interoperable communication protocols and realises a dynamic multimodal/heterogeneous network that can be deployed also in inaccessible, or remote spaces (oil platforms, mines, forests, tunnels, pipes, etc.) or in cases of emergencies or hazardous situations (earthquakes, fire, floods, radiation areas, etc.). In this infrastructure, these different entities or "things" discover and explore each other and learn to take advantage of each

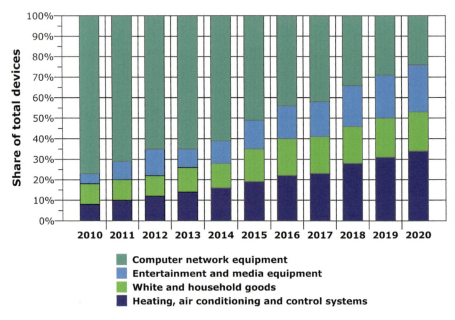

Fig. 2.4 Share of Internet-connectable consumer household devices by type, worldwide, 2010–2020 [1].

other's data by pooling of resources and dramatically enhancing the scope and reliability of the resulting services.

IoT is included by the US National Intelligence Council in the list of six "Disruptive Civil Technologies" with potential impacts on US [3]. NIC considers that "by 2025 Internet nodes may reside in everyday things — food packages, furniture, paper documents, and more." It describes future opportunities that will arise, starting from the idea that "popular demand combined with technology advances could drive widespread diffusion of an Internet of Things (IoT) that could, like the present Internet, contribute invaluably to economic development." The possible threats deriving from a widespread adoption of such a technology are also presented. It is discussed that "to the extent that everyday objects become information security risks, the IoT could distribute those risks far more widely than the Internet has to date."

The concept of Internet of Things is based on many enabling technologies that form the backbone of this new paradigm and for many people is rather abstract. In this context an interesting blog discussion started in 2010, that presents opinions on what the Internet of Things is not [6]. Based on the author's opinion the IoT is not:

- Ubiquitous/pervasive computing: Although the miniaturization of computing devices and the ubiquitous services derived from their data is probably a requirement for the IoT, pervasive computing is NOT the Internet of Things. Ubiquitous computing doesn't imply the use of objects, nor does it require an Internet infrastructure.
- The Internet Protocol: The Internet can be used globally because clients and servers use the same protocol for communication: however many objects in the Internet of Things will not be able to run an Internet Protocol.
- Communication technologies: As this represents only a partial functional requirement in the Internet of Things similar to the role of communication technology in the Internet and equalling communication technologies such as WiFi, Bluetooth, ZigBee, 6LoWPAN, ISA 100, WirelessHart/802.15.4, 18000-7, LTE to the Internet of Things is too simplistic. However, we can say that these technologies certainly might be part of Internet of Things.
- Embedded devices: RFID or wireless sensor networks (WSN), may be part of the Internet of Things, but as stand alone applications (intranets) they miss the back-end information infrastructures necessary to create new services. The IoT has come to mean much more that just networked RFID systems. While RFID systems have at least certain standardized information architectures to which all the Internet community could refer, global WSN infrastructures have not yet been standardized.
- Applications: A common misuse of the Internet of Things, very related with the pervasive computing issue and just as Google or Facebook could not be used in the early 90's to describe the possibilities offered by Internet or WWW. It is arguably to use Internet application and services to describe the Internet itself, but it is even more illogical to refer to small applications that would have no real impact on a global Internet.

In the vision of the Cluster these technologies are part of Internet of Things and are enablers of implementing the concept of Internet of Things in different applications. The IERC strategic research agenda is addressing

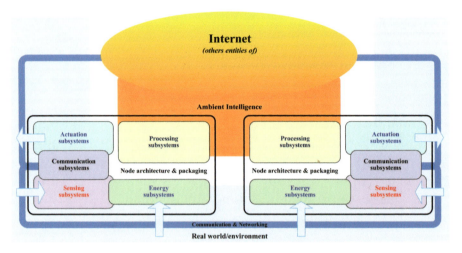

Fig. 2.5 Object connected to Internet of Things and their three main challenging domains: Technologies — Communication — Intelligence [15].

these challenges, considering and integrating the different point of views and differentiating between the Internet of Things from the other concepts and trying to identify the research needs for the implementation and deployment of IoT applications.

The interface between the real and digital worlds requires the capacity for the digital world to sense the real world and act on it. This implies the convergence of at least three domains: Technologies (nanoelectronics, sensors, actuators, embedded systems, cloud computing, software, etc.), Communication and Intelligence [15].

At the conceptual level the IoT technology represents the "middleware" between the implementation of the "grand challenges" such as climate change, energy efficiency, mobility, digital society, health at the global level and enabling technologies such as nanoelectronics, communications, sensors, smart phones, embedded systems, cloud computing and software technologies. These challenges will give rise to new products, new services, new interfaces and new applications. The "grand challenges" may also give rise to smart environments and smart spaces.

2.3 Technological Trends

Advances in wireless networking technology and the greater standardization of communications protocols make it possible to collect data from sensors and wireless identifiable devices almost anywhere at any time. Miniaturized silicon chips are designed with new capabilities, while costs, following the Moore's Law, are falling. Massive increases in storage and computing power, some of it available via cloud computing, make number crunching possible at very large scale and at a high volume, low cost.

It is possible to identify, for the years to come, a number of distinct macro-trends that will shape the future of ICT.

- First, the explosion in the volumes of data collected, exchanged and stored by IoT interconnected objects will require novel methods and mechanisms to find, fetch, and transmit data. This will not happen unless the energy required to operate these devices is dramatically decreased or we discover novel energy harvesting techniques. Today, many data centres have already reached their maximum level of energy consumption, and the acquisition of new devices can only follow the replacement of old ones, as it is not possible to increase energy consumption.
- Second, research is looking for ultra low power autonomic devices and systems from the tiniest smart dust to the huge data centres that will self-harvest the energy they need.
- Third, miniaturisation of devices is also taking place at a lightning speed, and the objective of a single-electron transistor, which seems to be (depending on new discoveries in physics) the ultimate limit, is getting closer.
- Fourth, the trend is towards the autonomous and responsible behaviour of resources. The ever growing complexity of systems, possibly including mobile devices, will be unmanageable, and will hamper the creation of new services and applications, unless the systems will show "self-*" functionality, such as self-management, self-healing and self-configuration.

The key to addressing these macro-trends by IoT is research and development, which drives the innovation cycle by exploiting the results to bring beneficial new technologies to the market and therefore into industrial applications.

IoT research and development is becoming more complex, due to the already highly advanced level of technology, the global, intersectoral and interdisciplinary collaboration needed and the ever increasing demands of society

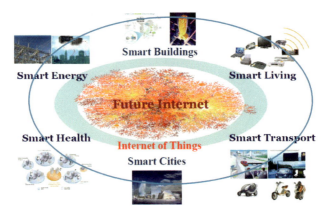

Fig. 2.6 IoT and smart environments creation.

and the economic global marketplace. Development of certain enabling technologies such as nanoelectronics, communications, sensors, smart phones, embedded systems, cloud computing and software technologies will be essential to support important future IoT product innovations affecting the different industrial sectors. In addition, systems and network infrastructure (Future Internet) are becoming critical due to the fast growth and advanced nature of communication services as well as the integration with the healthcare systems, transport, energy efficient buildings, smart grid, smart cities, and electric vehicles initiatives.

The focus of IoT research and development projects is on producing concrete results for several industries, which can then be further developed or exploited directly in creating smart environments/spaces and self-aware products/processes for the benefit of society.

2.4 IoT Applications

The major objectives for IoT are the creation of smart environments/spaces and self-aware things (for example: smart transport, products, cities, buildings, rural areas, energy, health, living, etc.) for climate, food, energy, mobility, digital society and health applications. The concept is illustrated in Figure 2.6.

The developments in smart entities will also encourage the development of the novel technologies needed to address the emerging challenges of public health, aging population, environmental protection and climate change, the conservation of energy and scarce materials, enhancements to safety and security and the continuation and growth of economic prosperity. These challenges will be addressed by:

- Providing reliable, intelligent, self-managed, context aware and adaptable network technology, network discovery, and network management.
- Refining the interaction between hardware, software, algorithms as well as the development of smart interfaces among things (smart machine to machine, things to things interfaces) and smart human-machine/things interfaces, thus enabling smart and mobile software.
- Embedding smart functionality through further developments in the area of nanoelectronics, sensors, actuators, antennas, storage, energy sources, embedded systems and sensor networks.
- Developments across disciplines to address the multi functional, multi-domain communications, information and signal processing technology, identification technology, and discovery and search engine technologies.
- Developing novel techniques and concepts to improve the existing security, privacy and business safety technologies in order to adapt to new technological and societal challenges.
- Enhancing standardisation, interoperability, validation and modularisation of the IoT technologies and solutions.
- Defining new governance principles that address the technology developments and allow for business development and free access to knowledge in line with global needs while maintaining respect for privacy, security and safety.

In this context Internet of Things applications are linked with the Green computing or Green ICT which is defined as "the study and practice of designing, manufacturing, using, and disposing of computers, servers, and associated subsystem — such as monitors, printers, storage devices, and networking and communications systems — efficiently and effectively with minimal or no impact on the environment" [8].

In the future most edge-connecting object-connected devices will have some form of wireless connectivity and the Internet of Things will drive energy efficient applications such as the power grid, or smart grid, connected electric vehicles, energy efficient buildings and will contribute to major savings in fuel consumption and hence carbon emissions. The Internet of Things technologies will allow greening of ICT by CO_2 reduction of infrastructure and products in ICT industry and greening by ICT applications by CO_2 reduction through convergence with ICT in other industries and industrial sectors. Internet of Things provides the technology and solutions that make full use of the integrated technologies of the communications networks and Internet technologies to build future oriented green intelligent cities, that provides a wide variety of interactive and control methods for the system of urban information and further support for building comprehensive systems for the development of urban ecology.

2.5 Technology Enablers

2.5.1 Energy

Energy issues, in all its phases, from harvesting to conservation and usage, are central to the development of the IoT. There is a need to research and develop solutions in this area (nanoelectronics, semiconductor, sensor technology, micro systems integration) having as an objective ultra low power devices, as current devices seem inadequate considering the processing power needed and energy limitations of the future. Using "More Than Moore's" technologies, that focus on system integration, will increase efficiency of current systems, and will provide a number of solutions for the future needs.

2.5.2 Intelligence

Capabilities such as self-awareness, context awareness and inter-machine communication are considered a high priority for the IoT. Integration of memory and processing power, and the ability to withstand harsh environments are also a high priority, as are the best possible security techniques. More specifically, the provision of security at physical layer, exploiting the characteristics of wireless channels, represents the envisioned low-complexity solution also addressing the scalability issues raised by large-scale deployments of smart "things". Transistor density is bound to grow, following Moore's Law, allowing therefore more "intelligent" electronics with increased on chip processing and memory capabilities. Novel cognitive approaches that leverage opportunistically on the time dependent available heterogeneous network resources can be adopted to support seamless continuous access to the information network as well as handle intermittent network connectivity in harsh and/or mobile environments. "Intelligent" approaches to knowledge discovery and device control will also be important research challenges.

2.5.3 Communication

New smart antennas (fractal antennas, adaptive antennas, receptive directional antennas, plasma antennas), that can be embedded in the objects and made of new materials are the communication means that will enable new advanced communications systems on chip, which when combined with new protocols optimized across the Physical (PHY), Media Access Control (MAC) and the Network (NWK) layers will enable the development of different Application Programming Interfaces (APIs) to be used for different applications. Modulation schemes, transmission rates, and transmission speed are also important issues to be tackled. New advanced solutions need to be defined to effectively support mobility of billions of smart things, possibly equipped with multiple heterogeneous network resources. Last but not least, network virtualisation techniques are key to the ensure an evolutionary path for the deployment of IoT applications with assured Quality of Service (QoS).

2.5.4 Integration

Integration of wireless identification technologies (like Radio Frequency Identification — RFID) into packaging, or, preferably, into products themselves will allow for significant cost savings, increased eco-friendliness of products and enable a new dimension of product self-awareness for the benefit of consumers. Integration requires addressing the need for heterogeneous systems that have sensing, acting, communication, cognitive, processing and adaptability features and includes sensors, actuators, nanoelectronics circuits, embedded systems, algorithms, and software embedded in things and objects.

2.5.5 Dependability

Dependability of IoT systems is of paramount importance; therefore the IoT network infrastructure must ensure reliability security and privacy by supporting individual authentication of billions of heterogeneous devices using heterogeneous communication technologies across different administrative domains. Reliable energy-efficient communication protocols must also be designed to ensure dependability.

2.5.6 Semantic Technologies and IoT

IoT requires devices and applications that can easily connect and exchange information in an ad-hoc fashion with other systems. This will require devices and services to express needs and capabilities in formalised ways. To facilitate the interoperability in the IoT further research into semantic technologies is needed. Examples of challenges are large-scale distributed ontologies, new approaches to semantic web services, rule engines and approaches for hybrid reasoning over large heterogeneous data and fact bases, semantic-based discovery of devices and semantically driven code generation for device interfaces.

2.5.7 Resource-constrained Scenarios for Business Based IoT

IoT implies that even the smallest device or sensor could be connected to the network. Research in wireless sensor networks has already resulted in promising solutions, tools and operating systems that can run on very small and resource-constrained devices. These solutions need to be evaluated in real large-scale industrial applications in order to illustrate business-based scenarios for IoT.

2.5.8 Modelling and Design

The design of large-scale IoT systems is challenging due to the large number of heterogeneous components involved and due to the complex iterations among devices introduced by cooperative and distributed approaches. To cope with this issue, innovative models and design frameworks need to be devised; for example, inspired by co-simulation methods for large systems of systems and hardware-in-the-loop approaches.

2.5.9 Validation and Interoperability

Standardisation is a must but it is not enough. It is a known fact that, even if following the same standard, two different devices might not be interoperable.

This is a major showstopper for wide adoption of IoT technologies. Due to the complex and diverse nature of IoT technologies only one interoperability solution may not be possible and integration is therefore required. Future tags and devices must integrate different communication schemes, allow different architectures, centralised or distributed, and be able to communicate with other networks. Interoperability of IoT technologies will always be a complex topic which requires research effort to address the new challenges raised. This for instance might be achieved by increased embedded intelligence and different radio access technologies sometimes even with cognitive capabilities. All these new emerging features together with the necessary intercommunication between different technologies will raise even more complexity in testing and validation and therefore common methodologies and approaches are necessary to validate and ensure interoperability in a coherent and cost effective way. The efforts necessary in achieving success in this area must not be underestimated as the results will serve to really exploit IoT research results by successful worldwide interoperable deployments. One of key success factor of GSM/UMTS/LTE technologies is that the specifications were developed together with the conformity and interoperability testing standards which included machine readable tests written in high level testing languages (like TTCN).

2.5.10 Standards

Clearly, open standards are key enablers for the success of wireless communication technologies (like RFID or GSM), and, in general, for any kind of Machine-to-Machine communication (M2M). Without global recognised standards (such as, the TCP/IP protocol suite or GSM/UMTS/LTE) the expansion of RFID and M2M solutions to the Internet of Things cannot reach a global scale. The need for faster setting of interoperable standards has been recognised an important element for IoT applications deployment. Clarification on the requirements for a unique global identification, naming and resolver is needed. Lack of convergence of the definition of common reference models, reference architecture for the Future Networks, Future Internet and IoT and integration of legacy systems and networks is a challanges that has to be addressed in the future.

2.5.11 Manufacturing

Last but certainly not least, manufacturing challenges must be convincingly solved. Costs must be lowered to less than one cent per passive RFID tag, and production must reach extremely high volumes, while the whole production process must have a very limited impact on the environment, be based on strategies for reuse and recycling considering the overall life-cycle of digital devices and other products that might be tagged or sensor-enabled.

2.6 Internet of Things Research Agenda, Timelines and Priorities

2.6.1 Identification Technology

Further research is needed in the development, convergence and interoperability of technologies for identification and authentication that can operate at a global scale. This includes the management of unique identities for physical objects and devices, and handling of multiple identifiers for people and locations and possible cross-referencing among different identifiers for the same entity and with associated authentication credentials. The IoT will include a very large number of nodes, each of which will produce content that should be retrievable by any authorized user regardless of its or if is a person of his/her position.

New effective addressing policies mobility management are required and frameworks are needed for reliable and consistent encoding and decoding of identifiers, irrespective of which data carrier technology that is used (e.g., whether linear or 2-D barcode, RFID, memory button or other technologies), including those that may be developed in the future. For some applications, it may be necessary to use encrypted identifiers and pseudonym schemes in order to protect privacy or ensure security. Identifiers play a critical role for retrieval of information from repositories and for lookup in global directory lookup services and discovery services, to discover the availability and find addresses of distributed resources.

It is vital that identification technology can support various existing and future identifier schemes and can also interoperate with identifier structures already used in the existing Internet and World Wide Web, such as Uniform Resource Identifiers (URIs).

Further research is needed in development of new technologies that address the global ID schemes, identity management, identity encoding/ encryption, pseudonymity, (revocable) anonymity, authentication of parties, repository management using identification, authentication and addressing schemes, and the creation of global directory lookup services and discovery services for Internet of Things applications with various unique identifier schemes.

2.6.2 Internet of Things Architecture Technology

The Internet of Things needs an open architecture to maximise interoperability among heterogeneous systems and distributed resources including providers and consumers of information and services, whether they be human beings, software, smart objects or devices. Architecture standards should consist of well-defined abstract data models, interfaces and protocols, together with concrete bindings to neutral technologies (such as XML, web services etc.) in order to support the widest possible variety of operating systems and programming languages.

The architecture should have well-defined and granular layers, in order to foster a competitive marketplace of solutions, without locking any users into using a monolithic stack from a single solution provider. Like the Internet, the IoT architecture should be designed to be resilient to disruption of the physical network and should also anticipate that many of the nodes will be mobile, may have intermittent connectivity and may use various communication protocols at different times to connect to the IoT.

IoT nodes may need to dynamically and autonomously form peer networks with other nodes, whether local or remote, and this should be supported through a decentralised, distributed approach to the architecture, with support for semantic search, discovery and peer networking. Anticipating the vast volumes of data that may be generated, it is important that the architecture also includes mechanisms for moving intelligence and capabilities for filtering, pattern recognition, machine learning and decision-making towards the very edges of the network to enable distributed and decentralised processing of the information, either close to where data is generated or remotely in the cloud. The architectural design will also need to enable the processing, routing, storage and retrieval of events and allow for disconnected operations (e.g., where network connectivity might only be intermittent). Effective caching, pre-positioning and synchronisation of requests, updates and data flows need to be an integral feature of the architecture. By developing and defining the architecture in terms of open standards, we can expect increased participation from solution providers of all sizes and a competitive marketplace that benefits end users.

In summary, the following issues have to be addressed:

- Distributed open architecture with end to end characteristics, interoperability of heterogeneous systems, neutral access, clear layering and resilience to physical network disruption.
- Decentralized autonomic architectures based on peering of nodes.
- Architectures moving intelligence at the very edge of the networks, up to users' terminals and things.
- Cloud computing technology, event-driven architectures, disconnected operations and synchronization.
- Use of market mechanisms for increased competition and participation.

2.6.3 Communication Technology

Billions of connected devices are pushing current communication technologies, networks and services approaches to their limits and require new technological investigations. Research is required in the field of Internet architecture evolution, wireless system access architectures, protocols, device technologies, service oriented architecture able to support dynamically changing environments, security and privacy. Research is required in the field of dedicated applications integrating these technologies within complete end-to-end systems.

In the Internet of Things the following topics related to communication technology have to be considered:

- Communication to enable information exchange between "smart things/objects" and gateways between those "smart things/objects" and Internet.
- Communication with sensors for capturing and representing the physical world in the digital world.
- Communication with actuators to perform actions in the physical world triggered in the digital world.

- Communication with distributed storage units for data collection from sensors, identification and tracking systems.
- Communication for interaction with humans in the physical world.
- Communication and processing to provide data mining and services.
- Communication for physical world localization and tracking.
- Communication for identification to provide unique physical object identification in the digital world.

In the IoT the range of connectivity options will increase exponentially and the challenges of scalability, interoperability and ensuring return on investment for network operators will remain.

In this context the communication needs will change and new radio and service architectures will be required to cater for the connectivity demands of emerging devices. The frequency spectrum allocation and spectrum masks will have to be adapted to the new bandwidth and channel requirements. New communications paradigms that use opportunistically the communication resources available at any given time will have to be adopted to provide seamless connectivity. Approaches based on the use of multiple radio bearers or inspired by cognitive radio technologies will have to be pursued to provide dependability, especially in harsh environments. Issues to be addressed:

Issues to be addressed:

- Internet of Things energy efficient communication multi frequency protocols, communication spectrum and frequency allocation.
- New efficient multiuser detection schemes.
- Software defined radios to remove need for hardware upgrades when new protocols emerge.
- Cognitive radio approaches tailored to low-power IoT devices
- Opportunistic communications paradigms
- Multi-radio wireless communications
- Reliable energy-efficient communication protocols to ensure dependability (e. g. in harsh environments)
- Connectionless communications, even beyond IP.
- High performance, scalable algorithms and protocols.

2.6.4 Network Technology

The evolution and pervasiveness of present communication technologies has promised to revolutionize the way humans interact with their environment. The Internet of Things is born from this vision in which objects form an integral part of the communication infrastructures that wire today's world. For this vision to be realized, the Internet of Things architecture needs to be built on top of a network structure that integrates wired and wireless technologies in a transparent and seamless way. Wireless network technologies have gained more focus due to their ability to provide unobtrusive wire-free communication. They have also become the leading area of research when combined with data collecting technologies used for environmental and object monitoring.

In this regard, wireless sensor networks promise low power, low cost object monitoring and networking, constituting a fundamental technology for the evolution towards a truly embedded and autonomous Internet of Things. Design objectives of the proposed solutions are energy efficiency, scalability since the number of nodes can be very high, reliability, and robustness and self healing.

Integration of sensing technologies into passive RFID tags allows new applications into the IoT context. Intel Labs [10] is involved in research and development focused on wireless identification and sensing platforms where the tags are powered and read by standard RFID readers, harvesting the power from the reader's querying signal. The wireless identification and sensing platforms have been used to measure quantities in a certain environment, such as light, temperature, acceleration, strain, and liquid level.

In Internet of Things scenarios, distributed (pre-)processing of sensor data is required in order to handle a massive amount of measurement data. Therefore, the communication network needs to support an intelligent distribution of (pre-)processing power, considering the status of the power supply of the involved (mostly embedded) devices, the energetic transport cost, as well as the processing power available at each device. The data exchange itself can be optimized by using and developing energy-efficient protocols that require a smaller number of bits to exchange for given information and that require less processing power for the data evaluation on a device level; an example for such a protocol is Binary XML [14].

Auto configuration and network service assembling is an important area since Internet of Things requires highly dynamic and flexible network domains. The handling of such systems is just feasible if the network configuration is automated and adaptable to the actual situation [14].

Research is needed on:

- Networks exploiting: On-chip technology considering on chip communication architectures for dynamic configurations design time parameterized architecture with a dynamic routing scheme and a variable number of allowed virtual connections at each output.
- Scalable communication infrastructure on chip to dynamically support the communication among circuit modules based on varying workloads and/or changing constraints.
- Power aware networks that turn on and off the links in response to bursts and dips of traffic on demand.
- Network virtualisation.
- Adaptability and evolvability to heterogeneous environments, content, context/situation, and application needs (vehicular, ambient/domestic, industrial, etc.).
- Solutions to effectively support mobility of billions of smart things
- Solutions to effectively support connectivity of (possible mobile) smart things equipped with multiple heterogeneous network resources
- Cross-cutting challenge covering Network foundation as well as Internet by and for People, Internet of Services, Internet of Contents and Knowledge, and Internet of Things.

IP provides today the protocol for implementing IoT applications. More research is required for IP technology and eventually the development of different post IP protocols optimized for IoT, compatible and interoperable with the exiting IP technologies.

Issues to be addressed:

- Network technologies (fixed, wireless, mobile etc.).
- Ad-hoc, wireless sensor networks.
- Autonomic computing and networking.
- Opportunistic networking

- Development of the infrastructure for "Network of Networks" capable of supporting dynamically small area and scale free connections and characteristics (typical social communities).
- Password and identity distribution mechanisms at the network level.
- Security and privacy of heterogeneous devices using heterogeneous communication technologies across different administrative domains.
- Anonymous networking.
- IP and post IP technologies.
- Traffic modelling and estimation to ensure efficient communication, load balance and end-to-end Quality of Service.
- Multipath, multi-constraint routing algorithms to enable load balancing among resources constrained intermediate nodes.

2.6.5 Software, Services and Algorithms

Only with appropriate software will it be possible that the Internet of Things comes to life as imagined, as an integral part of the Future Internet. It is through software that novel applications and interactions are realized, and that the network with all its resources, devices and distributed services becomes manageable. For manageability, the need for some sort of self-configuration and auto-recovery after failures is foreseen. The IoT is based on the coexistence of many heterogeneous set of things, which individually provide specific functions accessible through its communication protocol. The use of an abstraction layer capable of harmonizing the access to the different devices with a common language and procedure is a common trend in IoT applications. There are devices that offer discoverable web services on an IP network, while there are many other devices without such services that need the introduction of a wrapping layer, consisting of an interface and a communication sub-layers. The interface provides the web interface and is responsible for the management of all the in/out messaging operations involved in the communication with the real/physical world. The communication sub-layer implements the logic behind the web service methods and translates these methods into a set of device specific commands to communicate with the real/physical tagged objects.

Services play a key role: they provide a good way to encapsulate functionality (e.g., abstracting from underlying heterogeneous hardware or implementation details), they can be orchestrated to create new, higher-level functionality, and they can be deployed and executed in remote locations, in-situ on an embedded device if necessary. Such distribution execution of service logic, sometimes also called distributed intelligence, will be the key in order to deal with the expected scalability challenges. The middleware is defined as a software layer or a set of sub-layers interposed between the technological and the application levels. The middleware architectures proposed in many projects for the IoT often follow the Service Oriented Architecture (SOA) approach. The adoption of the SOA principles allows for decomposing complex systems into applications consisting of an ecosystem of simpler and with well defined components. The use of common interfaces and standard protocols is common in such systems. However typical SOAs fail to provide the loose coupling and proper separation between types and instances that are needed in domains that involve "things" (e.g. home automation). For instance two light appliances may offer the same type of service (turning light on and off) but different actual services, if only because they are located in different rooms. These loose coupling and proper separation between types and instances are however well known in Component Based Software Engineering (CBSE) approaches.

Tools to support the challenging design of large-scale IoT systems need to be developed as well. Such tools need to cope with the large number of heterogeneous components involved and with the complex iterations among devices introduced by cooperative and distributed approaches. Innovative models and design frameworks need to be devised to support such tools (e.g., inspired by co-simulation methods for large systems of systems and hardware-in-the-loop approaches).

Issues to be addressed include:
- Service discovery and composition.
- Service management.
- Object abstraction.
- Semantic interoperability, semantic sensor web etc.
- Data sharing, propagation and collaboration.
- Autonomous agents.
- Human-machine interaction.

- Self management techniques to overcome increasing complexities and save energy.
- Distributed self adaptive software for self optimization, self configuration, self healing.
- Lightweight and open middleware based on interacting components/modules abstracting resource and network functions.
- Energy efficient micro operating systems.
- Software for virtualisation.
- Service composition.
- Language for object interaction.
- Bio-inspired algorithms (e.g., self organization) and solutions based on game theory (to overcome the risks of tragedy of commons and reaction to malicious nodes).
- Algorithms for optimal assignment of resources in pervasive and dynamic environments.
- Modelling and design tools for IoT objects and systems
- Mathematical models and algorithms for inventory management, production scheduling, and data mining.

2.6.6 Cloud Computing

In its broadest form, a 'cloud' can be defined as "an elastic execution environment of resources involving multiple stakeholders and providing a metered service at multiple granularities for a specified level of quality (of service)." [9].

It is up to debate whether the Internet of Things is related to cloud systems at all: Whilst the Internet of Things will certainly have to deal with issues related to elasticity, reliability and data management etc., there is an implicit assumption that resources in cloud computing are of a type that can host and/or process data — in particular storage and processors that can form a computational unit (a virtual processing platform). However, specialised clouds may e.g., integrate dedicated sensors to provide enhanced capabilities and the issues related to reliability of data streams etc. are principally independent of the type of data source. Though sensors as yet do not pose essential scalability issues, metering of resources will already require some degree of sensor information integration into the cloud. Clouds may furthermore offer vital support to the Internet of Things, in order to deal with a flexible amount of data originating

2.6 Internet of Things Research Agenda, Timelines and Priorities

from the diversity of sensors and "smart things/objects." Similarly, cloud concepts for scalability and elasticity may be of interest for the Internet of Things in order to better cope with dynamically scaling data streams [9].

Deployment of 4G and other wireless broadband networks will support new cloud services, and demand for cloud services will drive network deployment. Additional utility services, such as voice recognition and other intelligent interfaces will become a part of cloud service platforms; and standards for cloud interoperability will develop so that data, applications, and environments can be ported between different cloud services [11].

Cloud computing is a building block of the Future Internet and it is expected that the Internet of Things will be the biggest consumer of Cloud. The IoT applications are composed of many detectors and services to manage them and are very dynamic involving rapidly varying data volumes and rates. Clouds provide an elastic facility to manage this variability. Of course a Cloud environment can also provide the services for analysis of the data streams often

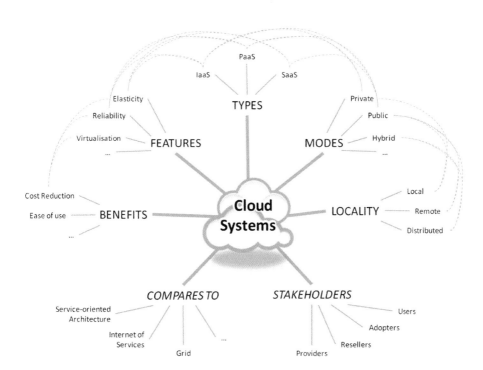

Fig. 2.7 Non-exhaustive view on the main aspects forming a cloud system [9].

associated with synchronous simulation to aid the provision of information to the end-user in an optimal form. The business benefit occurs in applications such as environmental monitoring, healthcare monitoring where the high volumes and rates of data need rapid processing to information for understanding.

2.6.7 Hardware

The developments in the area of IoT will require research for hardware adaptation and parallel processing in ultra low power multi processor system on chip that handle situations not predictable at design time with the capability of self-adaptiveness and self-organization. Research and development is needed in the area of very low power field-programmable gate array hardware where the configuration (or parts of it) is changed dynamically from time to time to introduce changes to the device. Context switching architectures, where a set of configurations are available and the device between switch between them depending on the defined using context.

Important issues are making a full interoperability of interconnected devices possible, providing the hardware with a sufficient degree of smartness by enabling their adaptation and autonomous behaviour, while guaranteeing trust, privacy, and security. In this context the IoT poses several new problems concerning the networking aspects when the things composing the IoT are defined in many cases by low resources in terms of both computation and energy capacity.

Research is needed for ultra low power very large scale integrated (VLSI) circuits containing scalable cognitive hardware systems that are changing the topology mapped on the chip using dedicated algorithms.

Self adaptive networks on chip that analyzes itself during run time and self adapts are required for IoT applications. Such run time adaptive network on chip will adapt the underlying interconnection infrastructure on demand in response to changing communication requirements imposed by an application and context.

Issues to be addressed:

- Nanotechnologies–miniaturization.
- Sensor technologies–embedded sensors, actuators.
- Solutions bridging nano and micro systems.
- Communication–antennas, energy efficient RF front ends.

- Nanoelectronics devices and technologies, self configuration, self optimization, self healing circuit architectures.
- Polymer electronics.
- Embedded systems–micro energy microprocessors/microcontrollers, hardware acceleration.
- Spintronics.
- Low cost, high performance secure identification/authentication devices.
- Low cost manufacturing techniques.
- Tamper-resistant technology, side-channel aware designs.

2.6.8 Data and Signal Processing Technology

By 2020, trillions of networked sensors will be deployed around the planet, in the spaces we inhabit, the systems we use, the devices we carry, and inside our bodies. Sensors are a key enabling technology; with detection, measurement, computation, and communication, they can make passive systems active. Sensors will be used to measure everything from acceleration and location to temperature, energy use, soil chemistry, air pollution, and health conditions. They will help ensure the structural integrity of airplanes, bridges, buildings and other critical infrastructure, and make our living environments more responsive to us. The streams of data they generate will support better management of resources and provide early warnings of significant events, from impending heart attacks to climate change. They will smooth transactions, and increase the visibility and transparency of previously obscure relationships and hidden economies. The information they provide will be actionable, and ultimately, provide us with greater foreknowledge and awareness of things to come [11].

In the context of Internet of Things the devices that are operating at the edge are evolving from embedded systems to cyber physical and web enabled "smart things/objects" that are integrating computation, physical and cognitive processes. Cognitive devices, embedded computers and networks will monitor and control the physical processes, with feedback loops where physical processes affect computations and cognitive processes and contrariwise. This convergence of physical computing and cognitive devices (wireless sensor networks, mobile phones, embedded systems, embedded computers, micro robots etc.) and the Internet will provide new design opportunities and challenges and requires new research that addresses the data and signal processing technology.

A typical features of cyber physical and web enabled "smart things/objects" will be the heterogeneity of device models, communication and cognitive capabilities. This heterogeneity concerns different execution models (synchronous, asynchronous, vs. timed and real-timed), communication models (synchronous vs. asynchronous), and scheduling of real time processes.

Issues to be addressed:

- Semantic interoperability,
- Service discovery,
- Service composition,
- Semantic sensor web,
- Data sharing, propagation and collaboration,
- Autonomous agents,
- Human machine interaction and human machine interfaces.

2.6.9 Discovery and Search Engine Technologies

The Internet of Things will consist of many distributed resources including sensors and actuators, as well as information sources and repositories. It will be necessary to develop technologies for searching and discovering such resources according to their capabilities (e.g., type of sensor/actuator/services offered), their location and/or the information they can provide (e.g., indexed by the unique IDs of objects, transactions etc.). Search and discovery services will be used not only by human operators but also by application software and autonomous smart objects, in order to help gathering complete sets of information from across many organisations and locations. Such services may also serve to discover what ambient infrastructure is available to support smart objects with their needs for transportation and handling, heating/cooling, network communication and data processing. These services play a key role in the mapping between real entities such as physical objects and in the assembly of their digital and virtual counterparts from a multitude of fragments of information owned and provided by different entities. Universal authentication mechanisms will be required, together with granular access control mechanisms that allow owners of resources to restrict who can discover their resources or the association between their resource and a specific entity, such as a uniquely identified physical object.

For efficient search and discovery, metadata and semantic tagging of information will be very important and there are significant challenges in ensuring that the large volumes of automatically generated information can be automatically and reliably accommodated without requiring human intervention. It will also be important that terrestrial mapping data is available and cross-referenced with logical locations such as postcodes and place names and that the search and discovery mechanisms are able to handle criteria involving location geometry concepts, such as spatial overlap and separation.

Issues to be addressed:

- Device discovery, distributed repositories
- Positioning and localisation
- Mapping of real, digital and virtual entities
- Terrestrial mapping data
- Semantic tagging and search
- Universal authentication mechanisms

2.6.10 Relationship Network Management Technologies

With many Internet of Things and applications moving to a distributed seamless architecture the future application manager needs to monitor more than just the infrastructure. The Internet of Things must incorporate traffic and congestion management. This will sense and manage information flows, detect overflow conditions and implement resource reservation for time-critical and life-critical data flows. The network management technologies will need depth visibility to the underlying seamless networks that serves the applications and services and check the processes that run on them, regardless of device, protocol, etc. This will require identifying sudden overloads in service response time and resolving solutions, monitoring IoT and web applications and identify any attacks by hackers, while getting connected remotely and managing all "smart things"/obejects involved in specific applications from remote "emergency" centres.

Issues to be addressed:

- Propagation of memes by things
- Identity, relationship and reputation management
- Traffic modelling and estimation

2.6.11 Power and Energy Storage Technologies

Objects require a digital "self" in order to be part of the Internet of Things. This participation is obtained by combining electronic identification, embedded and wireless communication technologies into the physical objects themselves. Simple digitalization alternatives, such as bar code and passive RFID, do not require an integral power source. More complex alternatives, such as those that provide active communications and object condition monitoring, need batteries to power the electronics.

Energy storage has become one of the most important obstacles to the miniaturization of electronic devices, and today's embedded wireless technologies such as Wireless Sensor Networks and Active RFID suffer from either bulky packaging to support large batteries or from short life times, that will require recharging or replacement of the integrated batteries. In order for the IoT to succeed in providing truly embedded and digital object participation, it is necessary to continue with the research on miniature high-capacity energy storage technologies. A solution that could bypass the shortcomings of energy storage is the harvesting of energy from the environment, which would automatically recharge small batteries contained in the objects.

Energy harvesting is still a very inefficient process that would require a large amount of research. Sources for energy harvesting in embedded devices could include, among others, vibration, solar radiation, thermal energy, etc.

Micro power technologies have emerged as a new technology area that can provide many development opportunities for IoT devices.

Research topics and issues that need to be addressed include:

- Energy harvesting/scavenging for MEMS devices and microsystems

- Electrostatic, piezoelectric and electromagnetic energy conversion schemes
- Thermoelectric systems and micro coolers
- Photovoltaic systems
- Micro fuel cells and micro reactors
- Micro combustion engines for power generation and propulsion
- Materials for energy applications
- Micro power ICs and transducers
- Micro battery technologies
- Energy storage and micro super capacitor technologies

2.6.12 Security and Privacy Technologies

Internet of Things needs to be built in such a way as to ensure an easy and safe user control. Consumers need confidence to fully embrace the Internet of Things in order to enjoy its potential benefits and avoid any risks to their security and privacy.

In the IoT every smart thing/object could be connected to the global Internet and is able to communicate with other smart objects, resulting in new security and privacy problems, e.g., confidentiality, authenticity, and integrity of data sensed and exchanged by 'things/objects'. Privacy of humans and things must be ensured to prevent unauthorized identification and tracking. In this context, the more autonomous and intelligent "things/smart objects" get, problems like the identity and privacy of things emerge, and accountability of things in their acting will have to be considered.

The close interaction of wirelessly interconnected things with the physical world makes it possible to pursue solutions that provide security at physical layer. Such solutions exploit the richness of the wireless channel features to ensure security at the physical layer. Their low-complexity solutions may help addressing at the same time scalability issues in large-scale IoT deployments.

The Internet of Things will challenge the traditional distributed database technology by addressing very large numbers of "things/objects" that handle data, in a global information space. This poses challenges. In this context the information map of the real world of interest is represented across billions of "things," many of which are updating in real-time and a transaction or data change is updated across hundreds or thousands of "things" with differing update policies, opens up for many security challenges and security techniques across multiple policies. In order to prevent the unauthorized use of private information, research is needed in the area of dynamic trust, security, and privacy management.

Issues to be addressed:

- Event-driven agents to enable an intelligent/self aware behaviour of networked devices
- Authentication and data integrity
- Privacy preserving technology for heterogeneous sets of devices
- Models for decentralised authentication and trust
- Energy efficient encryption and data protection technologies
- Security and trust for cloud computing
- Data ownership
- Legal and liability issues
- Repository data management
- Access and use rights, rules to share added value
- Responsibilities, liabilities
- Artificial immune systems solutions for IoT
- Secure, low cost devices
- Integration into, or connection to, privacy-preserving frameworks, with evaluation privacy-preserving effectiveness.
- Privacy Policies management
- Wireless security at physical layer

2.6.13 Standardisation

The Internet of Things will support interactions among many heterogeneous sources of data and many heterogeneous devices through the use of standard interfaces and data models to ensure a high degree of interoperability among diverse systems. Although many different standards may co-exist, the use of ontology based semantic standards will enable mapping and cross-referencing between them, in order to enable information exchange. From an architectural perspective, standards have an important role to play both within an organisation or entity and across organisations; adoption of standards promotes interoperability and allows each organisation or individual to benefit from a competitive marketplace of interoperable technology solutions from multiple providers; when those organisations or individuals which want to share or exchange information, standards allow them to do so efficiently, minimising ambiguity about the interpretation of the information they exchange. Standards regarding frequency spectrum allocation, radiation power levels and communication protocols will ensure that the Internet of Things co-operates with other users of the radio spectrum, including mobile telephony, broadcasting, emergency services etc. These can be expected to develop, as the Internet of Things increases in scale and reach and as additional radio spectrum becomes available through digital switchover etc.

As greater reliance is placed on the Internet of Things as the global infrastructure for generation and gathering of information, it will be essential to ensure that international quality and integrity standards are deployed and further developed, as necessary to ensure that the data can be trusted and also traced to its original authentic sources. In this context a close collaboration among different standardisation Institutions and other world wide Interest Groups and Alliances is mandatory.

Issues to be addressed:

- IoT standardisation
- Ontology based semantic standards
- Spectrum energy communication protocols standards
- Standards for communication within and outside cloud
- International quality/integrity standards for data creation, data traceability

2.7 Future Technological Developments

Development	2011–2015	2015–2020	Beyond 2020
Identification Technology	• Unified framework for unique identifiers • Open framework for the IoT • URIs	• Identity management • Semantics • Privacy-awareness	• "Thing DNA" identifier
Internet of Things Architecture Technology	• IoT architectures developments • IoT architecture in the FI • Network of networks architectures • F-O-T platforms interoperability	• Adaptive, context based architectures • Self-* properties	• Cognitive architectures • Experiential architectures
Communication Technology	• Ultra low power chip sets • On chip antennas • Millimetre wave single chips • Ultra low power single chip radios • Ultra low power system on chip	• Wide spectrum and spectrum aware protocols	• Unified protocol over wide spectrum
Network Technology	• Self aware and self organizing networks • Sensor network location transparency • Delay tolerant networks • Storage networks and power networks • Hybrid networking technologies	• Network context awareness	• Network cognition • Self learning, self repairing networks
Software and algorithms	• Large scale, open semantic software modules • Composable algorithms • Next generation IoT-based social software • Next generation IoT-based enterprise applications	• Goal oriented software • Distributed intelligence, problem solving • Things-to-Things collaboration environments	• User oriented software • The invisible IoT • Easy-to-deploy IoT sw • Things-to-Humans collaboration • IoT 4 All

(*Continued*)

2.6 Internet of Things Research Agenda, Timelines and Priorities

Development	2011–2015	2015–2020	Beyond 2020
Hardware	• Multi protocol, multi standards readers • More sensors and actuators • Secure, low-cost tags (e.g., Silent Tags) • NFC in mobile phones. Sensor integration with NFC	• Smart sensors (bio-chemical) • More sensors and actuators (tiny sensors)	• Nano-technology and new materials
Data and Signal Processing Technology	• Energy, frequency spectrum aware data processing, • Data processing context adaptable	• Context aware data processing and data responses	• Cognitive processing and optimisation
Discovery and Search Engine Technologies	• Distributed registries, search and discovery mechanisms • Semantic discovery of sensors and sensor data	• Automatic route tagging and identification management centres	• Cognitive search engines • Autonomous search engines
Power and Energy Storage Technologies	• Energy harvesting (energy conversion, photovoltaic) • Printed batteries • Long range wireless power	• Energy harvesting (biological, chemical, induction) • Power generation in harsh environments • Energy recycling • Wireless power	• Biodegradable batteries • Nano-power processing unit
Security and Privacy Technologies	• User centric context-aware privacy and privacy policies • Privacy aware data processing • Virtualisation and anonymisation	• Security and privacy profiles selection based on security and privacy needs • Privacy needs automatic evaluation • Context centric security	• Self adaptive security mechanisms and protocols
Material Technology	• SiC, GaN • Silicon • Improved/new semiconductor manufacturing processes/technologies for higher temperature ranges	• Diamond	
Standardisation	• IoT standardisation • M2M standardisation • Interoperability profiles	• Standards for cross interoperability with heterogeneous networks	• Standards for automatic communication protocols

2.8 Internet of Things Research Needs

Research Needs	2011–2015	2015–2020	Beyond 2020
Identification Technology	• Convergence of IP and IDs and addressing scheme • Unique ID • Multiple IDs for specific cases • Extend the ID concept (more than ID number) • Electro Magnetic Identification — EMID	• Beyond EMID	• Multi methods-one ID
IoT Architecture	• Extranet (Extranet of Things) (partner to partner applications, basic interoperability, billions-of-things)	• Internet (Internet of Things) (global scale applications, global interoperability, many trillions of things)	
SOA Software Services for IoT	• Composed IoT services (IoT Services composed of other Services, single domain, single administrative entity)	• Process IoT services (IoT Services implementing whole processes, multi/ cross domain, multi administrative entities, totally heterogeneous service infrastructures)	
Internet of Things Architecture Technology	• Adaptation of symmetric encryption and public key algorithms from active tags into passive tags • Universal authentication of objects • Graceful recovery of tags following power loss • More memory • Less energy consumption • 3-D real time location/position embedded systems • IoT Governance scheme	• Code in tags to be executed in the tag or in trusted readers. • Global applications • Adaptive coverage • Object intelligence • Context awareness	• Intelligent and collaborative functions
Communication Technology	• Long range (higher frequencies — tenth of GHz)	• On chip networks and multi standard RF architectures	• Self configuring, protocol seamless networks

(Continued)

2.6 Internet of Things Research Agenda, Timelines and Priorities

Research Needs	2011–2015	2015–2020	Beyond 2020
	• Protocols for interoperability • Protocols that make tags resilient to power interruption and fault induction. • Collision-resistant algorithms	• Plug and play tags • Self repairing tags	
Network Technology	• Grid/Cloud network • Hybrid networks • Ad hoc network formation • Self organising wireless mesh networks • Multi authentication • Sensor RFID-based systems • Networked RFID-based systems — interface with other networks — hybrid systems/networks	• Service based network • Integrated/universal authentication • Brokering of data through market mechanisms	• Need based network • Internet of **Everything** • Robust security based on a combination of ID metrics • Autonomous systems for non stop information technology service

Software and algorithms	• Self management and control
	• Micro operating systems
	• Context aware business event generation
	• Interoperable ontologies of business events
	• Scalable autonomous software
	• Software for coordinated emergence
	• (Enhanced) Probabilistic and non-probabilistic track and trace algorithms, run directly by individual "things."
	• Software and data distribution systems
		• Evolving software
		• Self reusable software
		• Autonomous things:
		   ○ Self configurable
		   ○ Self healing
		   ○ Self management
		• Platform for object intelligence
			• Self generating "molecular" software
			• Context aware software

Hardware Devices	• Paper thin electronic display with RFID • Ultra low power EPROM/FRAM	• Polymer based memory • Molecular sensors	• Biodegradable antennas • Autonomous "bee" type devices

(*Continued*)

Research Needs	2011–2015	2015–2020	Beyond 2020
	• NEMS • Polymer electronics tags • Antennas on chip • Coil on chip • Ultra low power circuits • Electronic paper • Devices capable of tolerating harsh environments (extreme temperature variation, vibration and shocks conditions and contact with different chemical substances) • Nano power processing units • Silent Tags • Biodegradable antennae	• Autonomous circuits. • Transparent displays • Interacting tags • Collaborative tags • Heterogeneous integration • Self powering sensors • Low cost modular devices	
Hardware Systems, Circuits and Architectures	• Multi protocol front ends • Multi standard mobile readers • Extended range of tags and readers • Transmission speed • Distributed control and databases • Multi-band, multi-mode wireless sensor architectures • Smart systems on tags with sensing and actuating capabilities (temperature, pressure, humidity, display, keypads, actuators, etc.) • Ultra low power chip sets to increase operational range (passive tags) and increased energy life (semi passive, active tags). • Ultra low cost chips with security • Collision free air to air protocol	• Adaptive architectures • Reconfigurable wireless systems • Changing and adapting functionalities to the environments • Micro readers with multi standard protocols for reading sensor and actuator data • Distributed memory and processing • Low cost modular devices	• Heterogeneous architectures • "Fluid" systems, continuously changing and adapting
Data and Signal Processing Technology	• Common sensor ontologies (cross domain)	• Autonomous computing	• Cognitive computing

(Continued)

2.6 Internet of Things Research Agenda, Timelines and Priorities

Research Needs	2011–2015	2015–2020	Beyond 2020
	• Distributed energy efficient data processing	• Tera scale computing	
Discovery and Search Engine Technologies	• Scalable Discovery services for connecting things with services while respecting security, privacy and confidentiality • "Search Engine" for Things • IoT Browser • Multiple identities per object	• On demand service discovery/integration • Universal authentication	• Cognitive registries
Power and Energy Storage Technologies	• Printed batteries • Photovoltaic cells • Super capacitors • Energy conversion devices • Grid power generation • Multiple power sources	• Paper based batteries • Wireless power everywhere, anytime. • Power generation for harsh environments	• Biodegradable batteries
Security and Privacy Technologies	• Adaptation of symmetric encryption and public key algorithms from active tags into passive tags • Low cost, secure and high performance identification/authentication devices	• Context based security activation algorithms • Service triggered security • Context-aware devices • Object intelligence	• Cognitive security systems
Material Technology	• Carbon • Conducting Polymers and semiconducting polymers and molecules • Conductive ink • Flexible substrates • Modular manufacturing techniques	• Carbon nanotube	
Standardisation	• Privacy and security cantered standards • Adoption of standards for "intelligent" IoT devices • Language for object interaction	• Dynamic standards • Adoption of standards for interacting devices	• Evolutionary standards • Adoption of standards for personalised devices

Acknowledgments

The IoT European Research Cluster — European Research Cluster on the Internet of Things (IERC) maintains its Strategic Research Agenda (SRA), taking into account its experiences and the results from the ongoing exchange among European and international experts.

The present document builds on the 2009 Strategic Research Agenda and presents the research fields and an updated roadmap on future R&D until 2015 and beyond 2020.

The IoT European Research Cluster SRA is part of a continuous IoT community dialogue initiated by the European Commission (EC) DG INFSO-D4 Unit for the European and international IoT stakeholders. The result is a lively document that is updated every year with expert feedback from ongoing and future projects within the FP7 Framework Program on Research and Development in Europe.

Many colleagues have assisted over the last few years with their views on the Internet of Things strategic research agenda document. Their *contributions* are gratefully acknowledged.

Ali Rezafard, IE, Afilias, EPCglobal Data Discovery JRG
Andras Vilmos, HU, Safepay, StoLPaN
Anthony Furness, UK, AIDC Global Ltd & AIM UK, CASAGRAS, RACE networkRFID
Antonio Manzalini, IT, Telecom Italia, CASCADAS
Carlo Maria Medaglia, IT, University of Rome 'Sapienza', IoT-A
Claudio Pastrone, IT, Istituto Superiore Mario Boella, Pervasive Technologies Research Area, ebbits
Daniel Thiemert, UK, University of Reading, HYDRA
David Simplot-Ryl, FR, INRIA/ERCIM, ASPIRE
Dimitris Kiritsis, CH, EPFL, IMS2020
Florent Frederix, EU, EC, EC
Franck Le Gall, FR, Inno, WALTER
Frederic Thiesse, CH, University of St. Gallen, Auto-ID Lab
Harald Vogt, DE, SAP, SToP
Harald Sundmaeker, DE, ATB GmbH, CuteLoop
Humberto Moran, UK, Friendly Technologies, PEARS Feasibility
John Soldatos, GR, Athens Information Technology, ASPIRE

Karel Wouters, BE, K.U. Leuven, PrimeLife
Kostas Kalaboukas, GR, SingularLogic, EURIDICE
Mario Hoffmann, DE, Fraunhofer-Institute SIT, HYDRA
Mark Harrison, UK, University of Cambridge, Auto-ID Lab, BRIDGE, EPC-global Data Discovery JRG
Markus Eisenhauer, DE, Fraunhofer-FIT, HYDRA, ebbits
Maurizio Spirito, IT, Istituto Superiore Mario Boella, Pervasive Technologies Research Area, ebbits
Maurizio Tomasella, UK, University of Cambridge, Auto-ID Lab, SMART, BRIDGE, Auto-ID Lab
Neeli Prasad, DK, CTIF, University of Aalborg, ASPIRE
Paolo Paganelli, IT, Insiel, EURIDICE
Philippe Cousin, FR, easy global market, Walter, Myfire, Mosquito, EU-China IoT
Stephan Haller, CH, SAP, CoBIS
Wang Wenfeng, CN, CESI/MIIT, CASAGRAS
Zsolt Kemeny, HU, Hungarian Academy of Sciences, TraSer

Contributing Projects and Initiatives
   ASPIRE, BRIDGE, CASCADAS, CONFIDENCE, CuteLoop, DACAR, ebbits, ETP, EPoSS, EU-IFM, EURIDICE, GRIFS, HYDRA, IMS2020, Indisputable Key, iSURF, LEAPFROG, PEARS Feasibility, PrimeLife, RACE networkRFID, SMART, StoLPaN, SToP, TraSer, WALTER, IOT-A, IOT@Work, ELLIOT, SPRINT, NEFFICS, IOT-I, CASAGRAS2, eDiana.

References

[1] Vision and Challenges for Realising the Internet of Things, European Union 2010, ISBN 9789279150883.
[2] Internet 3.0: The Internet of Things. © Analysys Mason Limited 2010.
[3] National Intelligence Council, Disruptive Civil Technologies — Six Technologies with Potential Impacts on US Interests Out to 2025 — Conference Report CR 2008–07, April 2008, Online: www.dni.gov/nic/NIC_home.html.
[4] ITU Internet Reports, The Internet of Things, November 2005.
[5] A Digital Agenda for Europe, COM (2010) 245, Chapter 2.5.3. Industry-led initiatives for open innovation.
[6] Extracting Value From the Massively Connected World of 2015, Online: www.gartner.com/DisplayDocument?id=476440.

[7] What the Internet of Things is NOT, Online: Technicaltoplus.blogspot.com/2010/03/what-internet-of-things-is-not.html.
[8] S. Murugesan, "Harnessing Green IT: Principles and Practices," IEEE IT Professional, pp. 24–33, January–February 2008.
[9] The future of Cloud Computing, Opportunities for European Cloud Computing beyond 2010 Online: cordis.europa.eu/fp7/ict/ssai/events-20100126-cloud-computing_en.html.
[10] Wireless identification and sensing platform, Online: seattle.intel-research.net/wisp/.
[11] ICT 2020_4 Scenario Stories. Hidden Assumptions and Future Challenges. Ministry of Economic Affairs, The Hague, 2010.
[12] G. Kortuem, F. Kawsar, V. Sundramoorthy, and D. Fitton, "Smart objects as building blocks for the internet of things," *IEEE Internet Computing* pp. 30–37, January/February 2010.
[13] Future internet 2020, Visions of an Industry Expert Group, May 2009.
[14] Future Internet Strategic Research Agenda, Version 1.1, January 2010.
[15] White Paper: Smart Networked Objects & Internet of Things, Les Instituts Carnot, V1.1, January 2011.

3

The Internet of Things: The Way Ahead

Dr. Gérald Santucci

European Commission, Belgium

3.1 Little Stones on the Road...

> "There are 35, roughly, billion devices on the network today — we'll have soon trillions of devices on the network that could free people to focus their energies on pressing issues like climate change or resource shortages."
>
> Dave Evans (Cisco Futurist), 2010

> "Can we really trust those companies which propose to us to shelter our data? Will they still exist in 10, 20, 100 years? How to make that they won't abuse valuable information, which would be detrimental to our individual freedom and our privacy? Is it the role of governments to provide these services? At a time when information spreads within a few minutes across the planet, to win — but above all, not to lose — the trust of the users is poised to become a crucial economic challenge. The world-wide computer and its thousands of interfaces allow us to catch a glimpse of the future nature of economic wars, in which trust and reputation will be at the heart of all battles."
>
> Frédéric Kaplan, La Métamorphose des Objets, 2009

> *"The nature of the IoT asks for a heterogeneous and differentiated legal framework that adequately takes into account the globality, verticality, ubiquity and technicity of the IoT."*
>
> Rolf H. Weber, Internet of Thing — New security and privacy challenges, 2010

> *"All that was object now becomes system. All that was even elementary unit, including and above all the atom, now becomes system. (...) We've always treated systems like objects; we've got now to conceive objects like systems."*
>
> Edgar Morin, La Méthode — 1. La Nature de la Nature, 1977

> *"It is only in the world of objects that we have time and space and selves."*
>
> T. S. Eliot

> *"I paint objects as I think them, not as I see them."*
>
> Picasso

> *"Ah, but a man's reach should exceed his grasp, or what's a heaven for?"*
>
> Robert Browning

3.2 The State of the Art of Internet of Things Research in Europe

3.2.1 The European Research Cluster on the Internet of Things

The European Research Cluster on the Internet of Things (IERC) [1] is a powerful instrument for bringing together the projects that address IoT research and innovation, thus enabling them to profit from each other's knowledge and experience, coordinating and encouraging the convergence of on-going work on the most important issues, and building a broadly-based consensus on the ways to realise IoT in Europe.

Reaching consensus is indeed essential to achieving widespread uptake of results in the field of IoT. Consensus involves not just technical agreement but also political acceptance and overall harmony between stakeholders. Standards have long been the basis of the world-wide development of any ICT. This continues to be the case in the emerging area referred to as "Internet of Things." The task of achieving consensus is becoming greater day-by-day requiring the deployment of dedicated resources and activities. The drive for consensus in defining IoT occurs within both traditional standards bodies (ISO/IEC, ITU, CEN, ETSI, ERO, etc.) and within the new industry for a (ZigBee Alliance, EPCglobal, AIM Global, NFC, IPv6 Forum, IPSO, etc.) These groups specifically dedicate work items to progress the definition of the IoT. Consensus is also the subject of specific work in IERC, within the framework of the FP7-ICT Challenge 1, and the European Commission (EC), within the framework of the Expert Group on the Internet of Things. IERC has identified a number of tasks having specific objectives to co-ordinate related activities in different areas. Projects responsible for these tasks are referred to broadly as "horizontal projects" (e.g., CASAGRAS2 and IoT-i). In general, horizontal projects set about defining common positions on topics of interest both for the projects, and for the development of IoT industry in Europe. The IERC is the instrument where "concertation" takes place: it develops the idea that "concertation" is an inherent feature of FP7-ICT Objective 1.3 at a number of different levels from policy/regulation to technological detail. The joint effort of all the projects in the IERC, with the support of the horizontal projects, can be viewed as progressing towards a co-operative system serving both the internal needs of IoT research and innovation projects and external needs.

Internal needs are met by plenary meetings of all Project Coordinators, which take place three or four times per year. In parallel to the plenary meetings, there may be meetings of relevant technology-oriented domains formed around the main technical areas as well as meetings of objective-driven chains, each supporting a defined objective and contributing to a specific result, acknowledged to be useful to the wider IoT community. External needs are met through the development of so called "guidelines" which can be seen as the collective output of the IERC to industry, governments and other important players who share in the vision of the IoT.

Therefore, "concertation" in the framework of clustering is a very important process for minimising overlaps while maximising synergies, supporting the attainment of key policy objectives, especially the relevant actions of the Digital Agenda for Europe (DAE) flagship, contributing to the overall Challenge 1 of FP7-ICT (i.e., the Future Internet), and of course addressing the main technological challenges.

The IERC includes more than 40 projects, the majority of them being finished RFID projects from FP6 and FP7 that have expressed the desire to remain involved in the evolution of the IERC as well as ongoing initiatives carried out at EU national level. The first set of "full-blooded" IoT related projects — eight projects — were launched in Autumn 2010 following the 5th call for proposals (Call 5) of the ICT Theme in the 7th Research Framework Programme (FP7).

Two "horizontal projects" exist at the present time: CASAGRAS2 and IoT-i. The goal of CASAGRAS2 (Coordination and Support Action for Global RFID-related Activities and Standardisation — 2) is to collaborate with IoT-I in forming an IoT Forum to address the key international issues that are important in providing the foundations and coordination necessary for realising the Internet of Things as a global objective. CASAGRAS2 examines features in the development of the IoT and work with disparate groups to achieve consensus with respect of integration and the realisation of the wider commitment to forming the IoT. The project outcomes will also serve as platforms for further individual and/or collective research towards the coordination of IoT development. The goal of IoT-i (Internet of Things Initiative) is to help establish a common vision of the Internet of Things, taking into consideration the large variety of requirements coming from the heterogeneous application domains. The vision is also founded on an understanding of the commonalities and peculiarities of existing and emerging technological solutions and their respective advantages and limitations. Furthermore, it has to be aligned with the vision of the larger Future Internet community in order to maximise the relevance and impact of the Internet of Things and its smooth integration. A dialogue has to be established between the currently disparate technology communities in Europe towards this vision and the research and development effort should be co-ordinated in such a manner that new synergies can emerge that are beneficial to all participating actors. The existing unification of the IoT research communities is not only a European issue but is applicable to all other continents. Establishing this unity and shared understanding in Europe is essential in gaining a competitive advantage over other countries.

3.2.2 Current European Research Projects

Although the IoT has been on the international policy agenda since several years (European Commission, IETF, ITU, OECD, etc.), collaborative R&D projects were not developed until 2010 when industry, academia and research centres in Europe formed consortia to respond to the Call 5 of FP7-ICT.

Below are presented the current R&D projects from FP7-ICT Call 5. Put all together, they represent an overall investment of 50 million Euro, including an EU funding of 33 million Euro.[1] The participants (77 in total) include Private Commercial Organisations (40%), Non-profit Research Organisations (21%), SMEs (18%), Higher or secondary Education Establishments (18%), and Others (3%).

3.2.2.1 EBBITS

(Enabling business-based Internet of Things and Services — An Interoperability platform for a real-world populated Internet of Things domain).

EBBITS performs R&D in the following areas:

- Internet of Things Architecture Technology using Service oriented Architecture (SoA) for maximum interoperability between heterogeneous structures and with end to end characteristics;
- Communication Technologies with distributed discovery architecture and unique physical identification of loosely coupled objects;
- Scalable Network Technologies integrating wired and wireless technologies using structured P2P networking layers in a transparent and seamless way;
- Software and Services with goal oriented orchestration and support for semantic interoperability, context awareness, and distributed decision support including workflow management and business rules processing;
- Security and Privacy Technologies enabled for cloud computing with models for decentralised identification, authentication and trust;

---

[1] Including the NEFFICS project which belongs to the cluster on Future Internet Enterprise Systems (FInES) but covers some aspects of the IoT.

> **EBBITS Vision**
>
> A widely deployed platform and service concept for the Internet of Things and Services, with which:
>
> - Producers can integrate physical devices, systems and components directly into their optimising systems, i.e. managing workflows, people, processes, assets, data, information and knowledge, and turn them into useful, value-added business services or service components.
>
> - Producers can obtain interoperability between various subsystems in manufacturing environments across manufacturing cells, manufacturing lines and entire manufacturing plants, regardless of geographical location with the aim to support production and energy optimisation.
>
> - Producers can meet increasing consumer demands and regulatory requirements for authentication and traceability of their products by providing support for authentication and traceability through ubiquitous services integrated in wireless communication networks and existing smart home infrastructures.
>
> - Producers, in particular SME's, of components, devices and systems can easily and cost-effectively network their products with mainstream enterprise systems in order to support higher value-added, interoperable solutions in an open architecture.

Fig. 3.1 EBBITS vision.

- Enterprise Framework and business socio-economic performance in dynamic business constellations and with sustainable business models.

3.2.2.2 ELLIOT

(Experiential Living Labs for the Internet of Things).

ELLIOT pursues the following scientific and technological objectives:

- To study and develop a set of Knowledge-Social-Business (KSB) Experience Models to represent human behaviour in the presence of IoT scenarios. Such models will be based on an advanced dynamic multimedia ontology (i.e., an ontology where concepts are not just texts but audio/video content) where speeches, gestures, movements, face expressions could be interpreted as manifestations of a KSB experience. Experiential models span along several dimensions: the subject (who, a human being), the object (what — a

> **ELLIOT VISION**
>
> An open architecture and platform to allow Enterprise Environments sharing KSB (Knowledge, Social and Business) assets that are derived from the human-centred experience of the existing seeds of the Internet of Things.
>
> This open architecture and platform will enable:
>
> - A seamless, trustworthy and pervasive infrastructure to bridge the conceptual, applicative and technological gaps between the societal, the information and the physical spaces in enterprise environments.
>
> - An increased adoption of IoT technologies and Ambient Intelligence (AmI) services thanks to the direct involvement of user/citizen communities (viral adoption where users co-create their own future) in the R&D process.
>
> - A dramatic increase of the innovation capacities through the complementary Open Innovation paradigm and co-creation Living Lab approach.

Fig. 3.2 ELLIOT vision.

product/service, either real or virtual), the context (e.g., where — space and site location, when — time and memory of the past, with whom — alone or in a group community, how — with pleasure, anger, bother);

- to design and develop the Experiential Platform. The challenge is to develop interoperability models being able to model and extract the essence of the experience by different cross-dimensional situations: how can we compare/foresee the behaviour of the same subject, experiencing the same object but in different space-temporal-psychological contexts? Or how can we establish a link between different subjects experiencing synchronously the same object in a co-location environment? Or what can we foresee by putting the same subjects in the same context as in a past experience, but giving the different objects to experience? The Experiential Platform will be able to interpret IoT situations and to classify them according to the KSB models defined above;

- to identify, experiment and explore IoT oriented User Co-creation Tools and Techniques. Co-creation in an IoT oriented environment is akin to the co-creation processes of software development. Especially in the area of Open Source Software (OSS), user co-creation is very common. Transferring the experience of OSS

developments into an IoT oriented user co-creation process through serious gaming seems to be a promising approach to enhance the capabilities in this area and to accelerate take-up and adoption;
- to explore, experiment and validate the approach in several scenarios conducted in three different Living Labs, each implementing different kinds of KSB experiences. All Living Labs are composed of a physical space artefact (i.e., a building, a civil architecture, a laboratory, an urban or rural delimited zone), an information space architecture (i.e., an interoperability framework, a collaboration space, an enterprise service bus, a semantically enabled service architecture), and a societal space community (i.e., a group of individuals, being workers/employees/customers/ suppliers/partners/consumers/students/professors/professionals/ retirees/citizens/ageing/disabled/patients, etc.).

3.2.2.3 IoT-A

(Internet of Things Architecture).

IoT-A's overall technical objective is to create the architectural foundations of the Future Internet of Things, allowing seamless integration of heterogeneous IoT technologies into a coherent architecture and their federation with other systems of the Future Internet. In order to achieve this ambitious overall goal, IoT-A has identified a series of detailed scientific and technological objectives that will be addressed within the context of the project:

- To provide an architectural reference model for the interoperability of IoT systems, outlining principles and guidelines for the technical design of its protocols, interfaces and algorithms;
- to assess existing IoT protocol suites and derive mechanisms to achieve end-to-end interoperability for seamless communication between IoT devices;
- to develop modelling tools and a description language for goal-oriented IoT-aware (business) process interactions allowing expression of their dependencies for a variety of deployment models;

> **IoT-A Vision**
>
> IoT will expand the boundaries of today's Internet to encompass the physical world and enable identification, information gathering and modification of the context. The term "things" in IoT refers to real physical objects like a truck, a coffee machine, a pallet, or a house. Usually some sort of device is attached to such a real-world entity in order to provide information about it or its environment, to interact with it, or to generate real-world events about it.
>
> These devices are expected to outnumber the human population by up to three orders of magnitude. Thus, the development of an efficient means for communication and interaction will be a necessity. This will provide the basis for building services onto the information and actuations they offer. As the denotation IoT implies, such devices will be connected not only in disjoint "Intranets", but they will offer dynamic, ubiquitous interconnectivity. Furthermore, all these things and the services interacting with them will feature security and privacy even down to hardware level. With respect to the sheer number of these devices, efficient selection mechanisms will have to be employed when sifting through the overwhelming wealth of information the devices provide.
>
> Real-world knowledge will be available in real time and on demand to all users including real-world centred applications, business processes, and enterprise systems. Thanks to standardised protocols, the IoT will enable the re-use and sharing of knowledge about the physical world. For example, IoT will enable "sensor commons", to which companies, public entities, and private persons will effortless connect.
>
> Achieving the vision of a globally interconnected and interoperable Internet of Things requires a deeper understanding of the requirements beyond application-domain-specific boundaries. We need to holistically encompass available precursors of IoT technology and consider the ongoing developments in other technological areas of the Future Internet.

Fig. 3.3 IoT-A vision.

- to derive adaptive mechanisms for distributed orchestration of IoT resource interactions exposing "self-properties" in order to deal with the complex dynamics of real world environments;
- to holistically embed effective and efficient security and privacy mechanisms into IoT devices and the protocols and services they utilise;
- to develop a novel resolution infrastructure for the IoT, allowing scalable look up and discovery of IoT resources, entities of the real world and their associations. (This infrastructure will be able to resolve names and identities to addresses and locators used by

communication services, thereby enabling cross-layer communication between IoT resources, services and applications);
- to develop the components required for the IoT device platform on which a future Internet of Things will be based, providing a basis for the research community to build upon;
- to validate the architectural reference model against the derived requirements with the implementation of real life use cases that demonstrate the benefits of the developed solutions;
- to contribute to the dissemination and exploitation of the developed architectural foundations. (The success of an architecture depends not only on its technical merits but also on its adoption by the community at large.)

3.2.2.4 IoT@Work

(Internet of Things at Work).

The scientific and technological objectives of IoT@Work are the following:

- To decouple automation application/controller programming from network operation by (1) achieving decoupling of application planning from network planning and configuration and reducing the effects of process and application reconfiguration on the amount of manual planning required at the network level, and (2) developing networking mechanisms to deliver advanced communication services that fulfil application demands in terms of reliability, real-time, scalability, and security;
- to enable self-operation of communication network (Plug&Work) by (1) enabling Plug&Work at all levels by developing autonomous network configuration for factory and industrial automation networks, and (2) developing an IoT-based architecture for autonomous network configuration taking application semantics and work-flows into account when structuring and optimizing the network operation;
- to ensure resilience and security in running automation systems by (1) supporting adaptive and agile manufacturing scenario while

3.2 The State of the Art of Internet of Things Research in Europe

> **IoT@Work Vision**
>
> Enriching the IoT architecture with the protocols and mechanisms required for supporting the factory/process automation networking needs.
>
> New concepts and architectures, focusing on autonomic network management and semantic Internet of Things, will be investigated in order to meet the requirements of a Future Internet-enabled factory. The process of composing IoT applications on top of a factory floor, in a device Plug&Work manner, impacts several layers, including the lower-level infrastructure (devices, links, and networks).
>
> IP technologies will continue to penetrate the industry in what is known as an "IP-to-the-field" approach, where each and every device/thing will be integrated into a multi-functional IT infrastructure, operating at both office and factory levels.
>
> As distributed systems are moving from being tightly coupled, to being more loosely connected (e.g., through web services), we are witnessing a move away from composition based on the usual functional interfaces and towards composition that takes into account a number of additional metadata that describe non-functional capabilities and requirements (e.g., deadlines, bandwidth, etc.).
>
> The automation applications, which are targeted by this project, will be set-up in a network agnostic manner, by relying on advanced communication services, which provide appropriate resources that match the specification of the application's non-functional requirements. These advanced network services will incorporate different self-organizing techniques, found in industrial networking technologies (e.g. Industrial Ethernet and WLAN) and other Internet protocols that have not yet been used to their full potential in automation networks.
>
> Thus, the IoT architecture will enable the secure sharing of network resources and automation devices by different applications, through intelligent, self-managing, Plug&Work mechanisms.

Fig. 3.4 IoT@Work vision.

securing and protecting the reliability and resilience of running systems, and (2) integrating strong security mechanisms at the architectural level in order to avoid unauthorized access to and interference with the production process.

3.2.2.5 SPRINT

(Software Platform for Integration of Engineering and Things).

To achieve this vision, the main objective of the project is to create the Internet of Engineering by combining the Internet of Physical Devices, the

> **SPRINT Vision**
>
> The SPRINT Networked Engineering Environment, i.e., the Internet of System Engineering based on an Internet-like service infrastructure, is intended to provide a simple and straightforward blending of all "things" pertaining to a complex system: sensors, actuators, electro-mechanical components, processing elements, system design tools and their models and development teams into a common platform with highly flexible collaboration and interoperability capabilities to support widely geographically-distributed teams through the entire cycle of complex system development: design, integration, verification and deployment.

Fig. 3.5 SPRINT vision.

Internet of Model Elements and the Internet of Designers.

- Internet of Physical Devices, where physical components are connected via the Internet to facilitate distributed design, remote simulation and testing; this is a concept that requires the development of new methods to model and evaluate correctness of complex interactions among physical devices.
- Internet of Design Model Elements, where model elements, regardless of their origin (tool, location, organization), can be located, referenced, shared and used across the Internet. For early verification, we need also methods to interface Physical and Virtual Devices.
- Internet of Designers, where engineers no matter where they are located can collaborate and cooperate over the Internet in an efficient and productive manner as if they where on the same organization and physical location. These three pillars are logically connected.

The specific objectives are defined as:

- To integrate tools and models of the whole development cycle;
- to integrate physical devices in the design process;
- to base the integration on semantic contracts;
- to demonstrate feasibility of the approach by applying it in industrial case studies;
- to create an open specification of the Internet of Engineering.

3.2.3 Internet of Things Research Project Portfolio

The European Research Cluster on the Internet of Things (IERC) has identified a number of key challenges which the current R&D projects are addressing in some way (see Table 3.1). The "concertation" taking place within the cluster will allow achieving consensus on significant issues like IoT Architecture, Governance, Privacy, etc.

In February 2011, the European Research Cluster on the Internet of Things and the EPoSS European Technology Platform sent to Vice President Neelie Kroes the IERC Position Paper on Research Priorities for Framework Programme 8.[2] This 12-page document provides a comprehensive analysis of the challenge for IoT research (technological trends, enablers), identifies the main pillars of IoT research for meeting societal demands (excellence, innovation, market deployment), and develops a rationale for strengthening IoT research in FP8 and make it an essential component of the Digital Agenda.

3.2.4 Main Scientific and Technological Challenges and Opportunities

Current R&D activities on the Internet of Things, sustained by FP7 financial support, are focusing on the development and standardisation of new technologies to enable the "Future Internet." Following Call 5, the second set of EU-funded projects stemmed from Call 7 of FP7-ICT. This call that had an indicative budget of 30 million Euro for "Internet-connected Objects" (Objective 1.3 in Challenge 1), generated an impressive number of 84 proposals, including 12 Integrated Project (IP) proposals, 67 Specific Targeted Research Project (STReP) proposals, and 5 Coordination and Support Action (CSA) proposals.

Given the EU funding possibilities, it is clear that only a few proposals will be selected, leaving aside several high-quality proposals that will have no other opportunity than to wait for the next relevant call for proposals, i.e., not before 2013. However, it is interesting browsing on all the proposals in order to discover what European actors have in their carton-boxes today and, in fact, to acknowledge the variety, depth and interdependence of the visions

---
[2]Position Paper on Research Priorities for FP8, 14 December 2010, Version 10.0.

Table 3.1. Mapping of IoT projects to issue areas.

	EBBITS	ELLIOT	IoT-A	IoT@Work	SPRINT
Architecture approaches and models	An open, service oriented infrastructure	An open architecture and experiential platform	Creation of an Architectural Reference Model	"Plug & Work" self configuration and self organisation	An internet-inspired open, extensible and loosely coupled integration approach, based on the IBM's JAZZ Integration Architecture (JIA)
Naming & addressing; search & discovery	Physical addressing schemes; virtualisation of the device (e.g., on a network node)		A resolution infra-structure to deal with the heterogeneous identification, naming, and addressing schemes that can be found in the IoT	Based on P2P technologies, and especially DHTs (Distributed Hash Tables)	Unified addressing scheme + common elements to enable tools to create resources and to view (and translate) them into a local format
Governance models	Open governance system (a Service Orchestration Manager federates the execution of the distributed services)		Open governance and avoidance of lock-in	Governance in remote access (claims, access rights, policies, security)	

(Continued)

Table 3.1. (Continued)

	EBBITS	ELLIOT	IoT-A	IoT@Work	SPRINT
Service openness; interoperability	Interoperability of the platform	Interoperability models able to model and extract the essence of the experience by different cross-dimensional situations	From interoperability at the connectivity level up to and including the information in the services level	Standard open operation with multi-vendor interoperability	A new semantics-based interoperability and collaboration infrastructure for integrating different tools and models among different companies and teams across the Internet
Privacy and security	Security and Privacy Technologies enabled for cloud computing with models for decentralised identification, authentication and trust	Data-security and privacy issues are considered	Privacy implemented at all levels of the architecture with minimisation of data collection, transfer and storage	Security in automation networks, network reliability and resilience	
Application scenarios	Two field trials: — Manufacturing — Lifecycle Management in the Food Chain	Logistics Product Lifecycle Management Personalised Media Services Tourism Service Public Transport Green Watch Vehicles as Environmental Sensors	— Health & Home — Retail & Automation (Logistics)	Remote maintenance of factory equipment and devices	Novel applications: — Physical devices on-line monitoring — Global integrated testing facilities — "Laboratory as a service" over the Internet

(Continued)

Table 3.1. (Continued)

	EBBITS	ELLIOT	IoT-A	IoT@Work	SPRINT
Pre-normative and/or pre-regulatory research	Contribution to standards in production optimisation, factory automation, farm systems integration, food traceability, semantic technologies, wireless networks, and RFID technologies		Participation in relevant standardisation	Contribution to the definition of relevant standards, such as IEEE, ISA, IEC, PNO	Contribution to the promotion of global standards (W3C, OMG, OASIS, ETSI, ISO/IEC, ZigBee Alliance, etc.)

(*Source:* Gérald Santucci & Peter Friess,INFO/D4).

and aims. Some interesting themes include:

- To design a general-purpose, open architecture, suitable for Internet-connected Objects, encompassing devices of heterogeneous nature (mobiles, embedded devices, PDAs, RFID), with different capabilities (processing power, bandwidth, energy resources, etc.) and diverse communication schemes;
- to conceive a proactive open architecture that supports cooperating Internet-connected objects seamlessly integrated in dynamic business environments with end-to-end characteristics by using and enhancing the current reference architectures and technologies of key sector specific communities (sensitive cargo transportation, critical asset management in life sciences, personal health systems);
- to characterize mobility patterns of benefit to WSN-based applications and develop novel approaches to manage mobility in 6LoW-PANs, create "plug-and-play" sensors for mobile environments that include Internet connectivity, and exploit IPv6 connectivity and mobility in 6LoWPANs in order to allow mobile objects to discover each other, semantically describe their services and interact, in interoperable and uniform ways;
- to define an open networked architecture and design and develop adaptive software to support transparent connection and interoperability between Wireless Sensor Networks and cloud computing infrastructures;
- to provide object-related information in a context-dependent manner (depending on time, location, requester, owner, and other parameters) using a distributed and decentralised ("beyond ONS") architecture in order to ensure secure and privacy-aware information provision over the complete lifecycle and from the various owners of an item;
- to couple and store the geo-referenced information to its physical geographical location, making use of the locally available storage, including the mostly unused, massively distributed storage capacity of mobile devices;
- to develop a novel set of components acting as middleware between producers and consumers of information, including a specialised

marketplace, the necessary publish-subscribe functions, accounting for production and consumption, open communication APIs, and agents that can implement specific policies and shield users and applications from system details;
- to develop an effective architecture and a test-driven service creation environment for IoT enabled business processes in order to accelerate the introduction of new services by providing orchestration, self-management capable components, and abstraction of the heterogeneity of underlying technologies to ensure interoperability;
- to implement the world's first environmental social network, composed of billions of devices (self-powered sensors placed in trash bins, mobile devices owned by citizens, and Access Points belonging to individuals, companies or governments), which will allow individuals to securely identify themselves using their mobile devices at any recycling point, anywhere in the world;
- to enable intelligent machine-to-machine communication among the interconnecting entities in order to provide seamless dissemination of services based on satellite data to remote end-users without human intervention and services interruption over the Internet;
- to construct a logic bus which will capture business events generated in heterogeneous information systems and electronic devices present in companies, supplying a dashboard for predicting and anticipating future situations and, overall, for analytic decision making in real time.

These themes, which are the first response to the research needs identified by the European Research Cluster on the Internet of Things,[3] highlight the need to address a number of scientific and technical challenges, among which the most important are the following:

- The integration of smart autonomous interconnected objects under strong energy, sustainability and environmental constraints;
- the massive, secure, privacy-aware, dynamic and flexible networking of objects;

---

[3] Position Paper on Research Priorities for FP8, op. cit., page 5/12.

- the fusion of the data obtained from the sensors, network and service management, the distributed data treatment, and ambient intelligence.

It is worth noting that several IoT proposals today establish a link between IoT and cloud computing. This trend seems logical and likely to gain momentum in the upcoming months. For example, clouds may offer vital support to the IoT, in order to deal with a flexible amount of data originating from the diversity of sensors and objects; furthermore, cloud concepts for scalability and elasticity may be of interest for the IoT in order to better cope with dynamically scaling data streams.

It is also interesting to note the variety of "field trials," "pilot applications" or "use cases" that European consortia envisage for the validation of their proposed solutions. Validation choices show that the European industry is now committed to fully exploit the potential of IoT technologies in innovative applications of both commercial and public interest: Agriculture, (Augmented services for) Airport passengers, Ambient assisted living, (Critical) Asset management in life sciences, (Sensitive) Cargo transportation, Cold chain monitoring, Collaboration of ubiquitous resources during ephemeral events, (Supply chain management in) Construction industry, (Critical and dangerous) Driving situations, Early warning system for natural disasters (tsunamis), E-health/homecare/hospital environment/medical surgery lifecycle in hospitals/personal health systems/mobile cardiac patient monitoring/well-being, Environment/intelligent sustainability/eco-energy/"Green Habitat," Event management, Geo-referenced information, Harbour environment, Home automation, (Critical) Infrastructures: power grids/ transportation/oil/gas/renewable energy, Integration of sensors and emergency support systems during wildfires and flooding, Intelligent (Smart) buildings, Logistics, (Intelligent, user-oriented) Maintenance and repair of cars, (Intelligent) Traffic management, Marine transport, Mobility, Piracy detection and avoidance, Public transportation, Recycling, Renewable energy production and collective optimisation, Retail environment/"intelligent retail shop," Road tunnel, Self-management of resources for the provision of smart grids, (Mobile) Social networks, Supply chain management, Smart City/ "living technology showcases"/municipality services/ridesharing, Smart infrastructures, Smart metering, Smart office, Smart toys, Smart transportation, Tourism, Video surveillance sensor networks, Waste collection management, and Working environments (factories or warehouses).

In some EU Member States, significant investment is also being made and will enable standardised solutions in a way complementary to the work carried out under EU FP7 umbrella. Two examples are given hereafter.

- The German Federal Ministry of Economics and Technology (BMWi)[4] supports a dozen critical IoT applications (Smart Home, Networked Manufacturing, Life Cycle Performance, e-Health, etc.), which are considered from different perspectives (technological evolution, standards and interoperability, international aspects, business models, etc.);
- the French Government supports through the Nov@log Competitiveness Cluster on Logistics the TACITES project (Tag Authentication and Convergence for Internet of Things and Enhanced Security), which aims to develop fully secure traceability solutions for combating counterfeiting in logistic and supply chains.

The two horizontal projects, CASAGRAS2 and IoT-i, together with the RACE Thematic Network, are implementing various strategies to raise awareness about RFID and IoT across Europe and help develop EU-, national- and regional-level initiatives towards the completion of a single IoT market.

Achieving a single market for IoT in Europe is a prerequisite to sustainable economic growth, especially in the post-crisis era. IoT is poised to play a significant role in the evolution of the ICT sector, which is an important element as the latter is the largest investor in R&D and it drives a large part of technical change and innovation. The widespread adoption of the IoT will take time, but the timeline is advancing thanks to improvements in key areas such as Future Internet technology and applications, wireless networking technology, standardisation of communication protocols, silicon chips capabilities, and storage and computing power (including via cloud computing). The use of

---

[4] At CeBIT 2011, the BMWi granted an award to a consortium that developed a "SensorCloud," i.e., a central and highly scalable platform for interconnected sensors and steering applications. The "SensorCloud" is opening up new possibilities to record, store and process measurement data from a broad range of industries and fields of application — from environment to traffic, energy, production machines or mobility, thus making a significant contribution to the "Internet of things."

RFID in supply chains has begun in the mid-2000s and is now taking place at the item level. Mobile phones increasingly incorporate support from NFC (Near Field Communication) technology, even on the SIM card to warrant security of transactions; "mobile wallets" will be commonly used within the next 3–4 years. As internetworking spreads and the Internet becomes pervasive, ubiquitous positioning technology will provide by 2018 opportunities for "whereness" applications in indoor and outdoor-to-indoor environments, for management of scarce resources, health and well-being, "self-actualisation," etc. After 2020, the results of today's R&D projects such as those funded under FP7 will be massively exploited, notably in the form of intelligent software that receives and analyses — with or without human intervention — large sets of data from connected everyday objects. After 2025, the interconnection of nanoscale devices with existing communication networks and ultimately the Internet will offer new solutions for many applications in the biomedical, industrial and military fields as well as in consumer and industrial goods, thus opening the way to an "Internet of Nano-Things."

The EU should foster collaborative R&D and innovation not only regarding the Internet dimension of the IoT (the network) but also regarding the "things" (the objects). The notion of "thing/object" is poised to change in the future given recent ICT developments that blur the boundaries between physical and virtual space. What is required is the development of proper applicative design solutions that allow today's connected objects to become genuine "actors" in the IoT. Objects of the future will be endowed with the capabilities of sensing/interpreting information that is not preformatted or generated in a predetermined context, reacting to that information and making adequate decisions, learning from experience and updating their own behaviour, and interacting with other objects to forge a "consensus" that will produce a specific collective behaviour [2]. *"The next tipping point that's going to get us into the Internet of things is having these computing devices be co-conspirators with us. (. . .) Think of an iPhone that is really more of your co-conspirator, or the knowledge navigator on your hip. It's a companion."* [3]. The embryos of such a "co-conspiracy" exist already as we see, for example, the private sector creating Web sites and smartphone apps that reformat information produced by a government agency in ways that are helpful to consumers, workers and companies [4]. The long-term consequences of the *metamorphosis of objects* [5, 6], especially on democracy and the role of nation-states, and the new economics that will accompany it, are today difficult to predict but will undoubtedly be far-reaching: *"In the history of mankind there are very few examples of empires dismantling themselves peacefully and leaving structures for after, people stepping down from their own free will and recognising their role, knowing when they are relevant and when they get to be antithetical to their original purpose."*[5]

Technical feats will not happen only in Europe. Other regions of the world are also increasingly investing in the IoT field in order to foster innovation, increase productivity, and boost their economies. The realisation of the full potential of IoT will require sustained cooperation between Europe and these regions, in particular for exchanging information and best practice on the outcome of pilot projects, reaching the critical mass of R&D needed to accelerate major breakthroughs in the various IoT technology areas, developing global standards, and aligning regulatory regimes as much as possible. In this context, it is not surprising that the recent set of proposals submitted to FP7-ICT Call 7 involved a significant number of non-European organisations, mostly in Coordination and Support Action proposals, but also, and this was new, in true shared-cost R&D proposals.

3.3 Towards an EU Policy Framework

The European Commission Communication on the Internet of Things [7] identifies fourteen lines of action that all together should provide Europe with a clear policy framework for the IoT:

- Governance;
- Continuous monitoring of the privacy and the protection of personal data questions;
- The silence of the chips;
- Identification of emerging security risks;
- IoT as a vital resource to economy and society;
- Standards mandate;
- Research and development;
- Public-private partnership;
- Innovation and pilot projects;

---

[5] With the courtesy of Rob Van Kranenburg, Council on the Internet of Things.

3.3 Towards an EU Policy Framework 75

Table 3.2. Non-European participants in new IoT research proposals.

Type of Proposal	China	Japan	South Korea	USA	Other countries
IP	— Wuxi Sensing Industrial Research Center	— Yokosuka Telecom Research Park			Brazil India
STReP	— East China Normal University — Hong Kong University — Nankai University	— National University Corporation Hokkaido University — NICT	— KAIST Korea Institute of Construction Technology	— University of California Los Angeles	Brazil Mexico
CSA	— Beijing University of Posts and Telecommunications — China Electronics Standardisation Institute — Wuxi Sensing Industrial Research Center	— YRP UNL — University of Tokyo	— Electronics and Telecommunications Research Institute (ETRI)	— University of Texas Arlington	Argentina Brazil Chile Columbia India Indonesia Malaysia South Africa Thailand Tunisia

(*Source*: Gérald Santucci).

> **The IoT seen by OECD**
>
> "Beyond the current Internet, a set of new technologies, such as radio frequency identification (RFID) and location-based technologies are predicted to enable new innovative applications and cause the network to evolve into an "Internet of Things". In the longer term, small wireless sensor devices embedded in objects, equipment and facilities are likely to be integrated with the Internet through wireless networks that will enable interconnectivity anywhere and at anytime. The future users and capacities of technologies that bridge the physical and virtual worlds are expected both to bring economic benefits and raise new societal challenges.
>
> An "Internet of Things" is predicted to be able to help individuals in their daily tasks and enhance business processes, supply chain management and quality assurance. It will enable distance monitoring of ambient conditions (e.g., temperature, pressure) and be used in a myriad of new applications, in areas such as healthcare and environmental monitoring. However, concerns relating to the invisibility of data collection and to the ability to trace and profile individuals could be exacerbated if tags and readers become pervasive and are combined with sensors and networks."
>
> Policy Brief, OECD, June 2008

Fig. 3.6 The IoT seen by OECD.

- Institutional awareness;
- International dialogue;
- RFID in recycling lines;
- measuring the uptake;
- assessment of evolution.

The EC adopted on 10 August 2010 a Decision to create an Expert Group on the Internet of Things [8]. This Expert Group will address the main challenges identified in the Communication and advise the EC on possible regulatory measures to ponder.

Privacy appears to be the most critical issue to tackle, as confirmed by a recent report of OECD (see box below). However, the deployment of the IoT also requires urgent action on other issues.

3.3.1 Privacy

Directive 95/46/EC [9] is the reference text, at European level, on the protection of personal data. It sets up a regulatory framework which seeks to strike a balance between a high level of protection for the privacy of individuals and the free movement of personal data within the European Union. To do so, the Directive sets strict limits on the collection and use of personal data and demands that each Member State sets up an independent national body responsible for the protection of these data. In November 2010, the Commission has unveiled recommendations to revise the Directive, in particular to address the need to bring European laws up to date with the challenges raised by new technologies and globalisation.

The Internet of Things marks obviously a technological disruption that will impact privacy and data protection. In particular the possibility of machine-to-machine communications based on identification and location, will allow for new ways to process information, bringing automated information management to a next level. In this respect, three consequences of IoT deployment need to be highlighted:

- While traditionally information about identified or identifiable persons was collected directly through the data subject, and the processing of such information was limited by the existing management tools, the IoT inevitably enhances the possibility of information collection about persons or objects;
- the IoT bears the potential to facilitate and even multiply the exchange of information and the processing of the information in general, i.e., the collection, use, consultation capabilities, and transfer or access of information by a variety of parties in a context where geographical barriers are blurred;
- the IoT multiplies the possibilities of automated decisions, which in turn might create an impression of loss of control either by the data subject or the data controller or by controllers who are responsible for the processing of the personal information.

The main IoT related challenges concerning privacy and data protection are the following[6]:

- Individuals need to be in control;
- privacy by design, data minimisation, purpose limitation;

---
[6]*Source*: Andreas Krisch, European Digital Rights (EDRi), hearing on IoT at European Parliament, 17 March 2010.

- data protection enforcement needs to be improved;
- global data protection standards are needed.

Leaving aside the well-known concepts of Privacy Enhancing Technologies[7] (PETs), Privacy-by-Design[8] and Privacy-by-Default,[9] a number of new regulatory strategies have been recently envisaged to tackle privacy challenges, especially in the IoT:

- *The Right to be Forgotten*: this concept, born from the concept of *Right to Forget* which, at the beginning of 2010, was considered by France for becoming a law that would have allowed everyone to ask that any information about them be deleted after a certain period of time, has been introduced by the European Commission in the revision of the EU Data Protection Directive [10]. The origin of the concept can also be found in [11], where the author makes individuals aware of the timelessness of what they created online and of what endless accumulation might portend.
- *The Right to (Digital) Oblivion*: this concept, which has the same meaning as the concept of right to be forgotten, is commonly used in Canada where the issue of data retention has become prominent in the last few years with the extension of Canadian privacy legislation to cover the private sector. It refers to the legal restrictions on the retention of records to protect individuals from the unreasonably long retention of possibly harmful data.
- *The Right to the Silence of the Chips*: this concept was coined by Bernard Benhamou, Delegate on Internet Usage at French Ministry of Research, at the time of the French Presidency of the EU during the second half of 2008. It means that individuals should be empowered with a technological solution (software) enabling them to disconnect from their networked environment at any time. More concretely, any chip could be given an On/Off function which

---

[7]A posteriori actions, modifications or corrections of an existing technology, in response to identified privacy protection problems (*Source*: German Federal Office for Information Security).

[8]To identify and examine possible data protection problems when designing new technology and to incorporate privacy protection into the overall design (*Source*: German Federal Office for Information Security).

[9]Provided options and capabilities to protect the privacy are activated or given by default (*Source*: German Federal Office for Information Security).

would allow the end-user to determine whether he would prefer to have the additional services, like product recall, from a continuation of the functioning of the chip after the object was sold. Such a function would strengthen the individual right of informational self-determination of citizens. The term was introduced in the EC Communication on the Internet of Things.

The first two concepts, though interesting and deserving wide debate among stakeholders, may be difficult to implement. If the *right to be forgotten* means content created by a person asking for it to be deleted (like e-mails), it sounds feasible. But if it refers to any content about a person, for example embarrassing pictures which friends of that person put on line, say on Facebook, MySpace or Wikipedia, there are strong reasons to believe that no matter what law puts in place, the information will not be suppressed. The Internet never forgets! The concept is certainly interesting but its actual implementation raises some practical questions:

- First, how would users report inaccurate or libellous information? Do they send an e-mail to the webmaster? Do they file a report with law enforcement?
- Second, does the web publisher have the right to appeal the request? Where start libellous statements and where finishes legal speech? It might be up to the courts to decide whether the information was in fact libellous.

More fundamentally, the application of the right to be forgotten raises philosophical as well as legal questions. People who want to demand content to be removed from the Internet, just because they are in it, are asking their acquaintances to censor their thoughts, edit their memories. In other words, have we any right to demand the removal of content which is the memory of an event where different personal timelines have intersected? There will be legal experts to argue that a right to be forgotten is a clear restriction on free speech. Not only will they debate the balance between free speech and privacy rights, they will also debate the very nature of privacy. Shouldn't a right to privacy only concern information that is actually private? Doesn't a right to be forgotten concern information that is, by default, public information, and pretending that it's private?

As a consequence, it is becoming urgent that Europe works on a clear common definition of the term *privacy*. It may refer indeed to different yet interdependent notions: Respect for intimacy, confidentiality, private information, respect for the physical and moral integrity of individuals, legal protection of personal data, and freedom of expression and movement.

The third concept — right to the challenge of the chips — is of a slightly different nature. Where law alone cannot solve the problems posed by technology, why not turning to technology to bring the solutions to the problems it creates? In this respect, the right to the silence of the chips refers to a legal right that would be supported by a technological solution that does not exist today. More concretely, cryptography could be used as the basis of a solution for deactivating a tag — and reactivating it at will. Some companies are already working on such a solution to ensure privacy, for example in Denmark and in France.

In the specific and challenging context of the IoT, the principle of data minimisation can be served through the application, where appropriate, of anonymisation and pseudo-anonymisation techniques, including encryption of personal data, based on a thorough analysis of the personal data processing and the context in which it takes place. Since data controllers have to conform also to the principle of purpose limitation, the transfer or granting of access rights to processors or other controllers will require an explicit outline of the specific purposes for which the personal data is collected. In a context where more and better managed information will be available, the access to and processing of personal data will need to be subject to scrupulous compliance with the existing legal obligations.

Furthermore, as the IoT will be supported by infrastructures of a very different nature, many of which being based on the applications it serves, we can foresee centralised repositories or networked based applications, close and open systems, interconnected or isolated. The information captured can be represented in multiple ways, and therefore interoperability will be essential in order to allow for efficient compliance with access and other individual rights of the data subject. In addition, these new developments will require the creation of efficient and tailored identity management tools, based on identification and authentication of the data subject (in order to guarantee the exercise of his/her rights) and other parties that are legitimised and obliged to process information in accordance with the law. Moreover, access rights by

Table 3.3. ENISA recommendations on the use of IoT and RFID technology in an automated air travel scenario.

Policy recommendations	Research recommendations	Legal recommendations
Rethink existing business structures and introduce new business models	Data protection and privacy	Take all the necessary technical and organisational measures to ensure the security of the personal data of the data subjects
User-friendliness of devices and procedures/be inclusive	Usability	Develop guidelines on the better enforcement and application of the European regulatory framework
Raise awareness/educate specialised personnel and citizens	Proposing standards of light cryptography protocols	Adopt an end-to-end approach for securing IoT/RFID applications (RFID tags, smart devices, readers, back-end databases)
Develop and adopt policies for data management and protection	Managing trust	Promote the participation of industry, in particular SMEs, in EU research and innovation activities
	Multi-modal person authentication (biometrics)	Encourage research at EU level on the ethical limits of private data capture and circulation

(*Source*: ENISA, 2010).

data controllers and processors will need to be more sophisticated and complex as not all data controllers and processors have the right to access and process all information available. This is of particular relevance in a context where the existence of joint data controllers, rather than a unique data controller, is envisaged, especially in the emergence of legitimate applications that are based on collaborative relationships.

3.3.2 Security

In 2010, following up on EC Communication on the IoT ENISA — European Network and Information Security Agency — has analysed the risks associated with a future air travel scenario, enabled with IoT/RFID technology. The ENISA report [12] identifies major security risks, as well as privacy, social and legal implications and also makes concrete policy and research and legal recommendations. Although limited to the specific case of air transport, the report provides generic recommendations that should be considered for many other sectors and applications.

Two key aspects of security need to be distinguished in the IoT. On one hand, the reliability of IoT systems has to be ensured: neither the systems nor the processed data or the data processing may be jeopardised in their existence, usage and availability. On the other hand, IoT systems have to be controllable: rights or other legitimate interests of affected persons may not be jeopardised by their existence or usage.

To meet these aspects of IoT security, several components are required:

- *Confidentiality*: no unauthorised access to data;
- *Integrity*: no unauthorised/unrecognised manipulation of data;
- *Availability*: processing of the systems functions at the defined time, within the defined period of time;
- *Accountability*: on every function (and its results) of a system, it must be possible to determine which instance (and/or person) triggered its processing;
- *Liability*: on every function (and its results) of a system, it must be legally provable which instance (and/or person) is responsible for it.

The existing Internet and information infrastructures pose a number of security problems (e.g., SPAM, DoS attacks, identity theft, viruses) which require adequate solutions. IoT systems are confronted with the same problems but also with new ones which are likely to gain importance in the future.

- To maintain *confidentiality and integrity of data*, mechanisms will be required to restrict access to information stored on objects and control whether an object is permitted to participate (connect, transmit or receive information) in an IoT system in general or at any given time. While available infrastructures might always be used to transmit confidential information between objects and applications, not all information transmitted via an IoT system is meant to be public. Therefore, to avoid unauthorised access to and manipulation of the transmitted data, mechanisms for end-to-end encryption and digital signatures will be required;
- to ensure the *availability of services*, the interoperability of the participating systems will have to be ensured by establishing a set of publicly accessible standards. Mechanisms for addressing a variety

of heterogeneous objects and to discover information will need to be established and public unrestricted access to these services will need to be ensured by public authorities;
- while it is already difficult to determine the responsible parties for certain activities on the existing Internet, this task of ensuring *accountability and legal liability* will be even more difficult with the IoT as there will be an indefinite number of micro-systems participating in highly dynamic, constantly changing capillary networks.[10]

Another issue, which is very important, concerns the need for users to keep control of the IoT components which they possess. These components, be they ordinary objects, useful gadgets or gizmos equipped with computing power, will often lack any kind of user interface but will need to be managed via specialised control devices. To avoid that users lose control of their IoT equipment, which could result in insecure systems and/or technology paternalism, such devices will need to reduce the complexity of the overall system to a clear and understandable user interface that can be used even by inexperienced people.

A last issue is the link between security and privacy. To take just one example, the U.S. Federal Trade Commission (FTC) has recently proposed to expand the scope of "personally identifiable information" to include URL and IP addresses. However, to protect consumers, security companies process URL and IP addresses not to identify individuals but to understand which machines are propagating malware. This is a rapidly evolving field and a good understanding of new and innovative technologies is essential to better target regulation in a way that caters for new developments.

3.3.3 Ethics

As pioneered by the Ethical Social and Legal Aspects working group on biotechnology, established by the European Commission in the early 1990s, there is an increasing need for a systematic analysis of the ethical aspects of ICT research. The emergence of the IoT, which may eventually enable

---

[10] The IoT calls for a new class of network, called capillary networks. These short-distance edge networks extend existing networks and services to all devices equipped with sensors and actuators and the physical environment in general.

object-to-object communications without human intervention, dramatically stresses this need.

The IoT is obviously not the only domain of the ICT sector where a debate within the community of researchers, scientists and engineers should be organised. So far, the focus has been the use of ICT implants in the human body, in particular RFID and biometrics identification, allegedly for medical purposes. However, philosophical questions have been raised regarding the impact of ICT implants on the "self," due to the invasive nature of the surgery, and on personal identity itself.

The IoT does not concern objects only; it is about the relations between the everyday objects surrounding humans and humans themselves. Therefore, the reflection on IoT and Ethics should also involve the Civil Society — privacy groups, consumer organisations, legal experts, thinkers, designers, philosophers, etc. The launch of such a reflection is urgent in order to prevent the risk that it lags behind the technological developments that are already moving fast. It is necessary to think through the ethical issues before developments proceed rather than afterwards when science and industry may be constrained by regulations that stifle innovation and hamper just and balanced deployments.

The ethical worries are manifold:

> *"To what extent can surveillance be accepted? How much control should be delegated to machines? How much transparency is needed for machine operations? How should content be shared through systems? (...) Does sensitivity to privacy drive ethical acceptability? Security? Universal Usability? Control? What standards need to be met? What goal we are striving for when we debate what's ethical and what isn't?"*[11]

The challenge is paramount. Since the beginning of the 21st century the confines of personal life have imperceptibly, yet dramatically changed. The issue of IoT and Ethics is not only an issue about the behaviours of scientists, engineers and/or merchants, and their consequences for all citizens. It's above all an issue about ourselves — how we see our role and our place in today's

---

[11] *Source*: Sarah Spiekermann, *System Design and the 'Idea of Man'*, 2010.

society. An increasing number of younger generation people keep an objective record of their lifestyles: Food, location, mood, physical exercise, sex, sleep, productivity, spiritual well-being, etc. Instead of questioning their inner worlds through talking, reading and writing, they are constructing a "quantified self" by using numbers [13].

At the edge of this trend, there is, for example, Gordon Bell, principal researcher in Microsoft Research, who is putting all of his atom- and electron-based bits in his local Cyberspace — it is called by MyLifeBits. Bell has captured a lifetime's worth of articles, books, cards, CDs, letters, memos, papers, photos, pictures, presentations, home movies, videotaped lectures, voice recordings, phone calls, IM transcripts, television, and radio, and stored them digitally. He has constructed "a life time store of everything." What should we think about this? Where does it bring us?

On one hand, there is of course the admiration in front of what modern technology can provide — smaller and better electronic sensors, increasingly powerful computing and communicating devices that individuals can carry on them (and someday *in* them), social media making it easy to share everything, and cloud computing which, because of the unprecedented scale and heterogeneity of the required infrastructure, may eventually shelter a global super-intelligence.

On the other hand, the tremendous potential of modern technology, in particular IoT, interrogates all humans about the place of human life in a world dominated by objects, be they machines, robots, or whatever other "things." When objects become social actors, as the IoT lets us foresee, when they interact among them and with human beings in their everyday lives, when they handle tasks for human beings, in some cases autonomously, people expect these objects to "think" and "act" like people. This implies that all objects in the IoT should be technically designed in such a way that they observe human values and behavioural norms.

Beyond privacy-by-design, the next big challenge is probably "Ethics-by-Design." The latter concept implies, among many things, the right of individuals to privacy, the right for people to make autonomous decisions and control their networked environment, and a new paradigm of accountability and liability for the actions undertaken by objects in the name of human beings, or for them.

3.3.4 Governance

The debate on the governance of the IoT is just beginning in Europe, although one element of it (the Object Naming Service, ONS) gave rise in 2005 to informal discussions triggered by the European Commission.[12] Over the last few years, the concept and scope of IoT governance have significantly expanded. Most stakeholders today agree that there is no need to have a global governance mechanism for an ONS which is just a technical service linking numbers to an object to allow companies to streamline their businesses, reduce costs, and make services more efficient and effective. Clear general competition and anti-trust rules as well as interoperable standards should be sufficient. However, many experts[13] argue that Europe should initiate a global discussion how an adequate global governance system could be developed to, on one hand, learn the lessons from (Internet Corporation for Assigned Names and Numbers) ICANN and, on the other hand, avoid the emergence of market-driven new monopoly with unintended negative political, economic, commercial and social effects.

Whatever form it takes, governance is by nature a solution for a problem; therefore, identifying that problem properly is a prerequisite. Since the difference for the acceptance of a service in the market place is the critical mass of users the service can generate and the user trust, the EU should foster innovation in the development of attractive competitive services which meet the needs of large and small user groups. Any governance mechanism should be designed on the basis of a comprehensive understanding of the needs of users and the dynamics of the market for IoT services; it should also take into consideration the legitimate public policy concerns such as privacy, data protection, and data security.

---

[12] On 13 June 2005, the EC initiated a series of consultations on the governance of the IoT with a first meeting in Brussels, which involved a handful of well-recognised international experts — Kim Davies (CENTR), Sabine Dolderer (DENIC), Patrik Fälström (Cisco), Peter Janssen (DNS.be/EURid), and Daniel Karrenberg (RIPE NCC) — to debate the concerns raised in Europe, primarily by certain governments, concerning the possible domination by two interlinked non-European entities on the allocation of electronic product codes and the management of the database, thus being the central pointer to the content of the information being circulated. Over the period 2006–2008, this inaugural meeting was followed by wide consultations with both EU national governments (Berlin IoT conference, June 2007, Lisbon IoT conference, November 2007, Nice IoT conference, October 2008) and private stakeholders, including VeriSign and Afilias.

[13] For example: Rolf H. Weber (University of Zurich), Francis Muguet (University of Geneva), Andreas Pfitzmann (Technical University in Dresden).

As shown in the EC Communication on the Internet of Things, the governance of the IoT involves technical, legal, economic and political issues. When accessing information related to an object, we are confronted with a number of key issues:

- Object naming: how is this identification structured?
- Assigning: Who assigns the identifier?
- Addressing: How and where can additional information about an object be retrieved, including its history?
- Security: How is information security ensured?
- Accountability: Which stakeholders are accountable for each of the above questions, what is the accountability mechanism?
- Ethics: Which ethical and legal framework applies to the different stakeholders?

IoT systems which have not properly addressed these questions could have serious negative implications, such as:

- Mishandled information could reveal an individual's personal data or compromise the confidentiality of business data;
- unsuitable assignation of rights and duties of private actors could stifle innovation;
- lack of accountability could jeopardise the functioning of the IoT system itself.

Some authors go even further by considering a wider scope for IoT governance.

As regards the governance of the IoT network infrastructure[14], there seems to be an agreement in Europe that a compromise must be found between an innovative environment and economic interests, on one hand, and societal and political interests, on the other hand. A governance model should allow for a minimum of specific regulation yet maintain the option for public authorities to react adequately in order to prevent monopolies and promote a dynamic market for IoT in the future. The minimum set of public requirements could turn around the following elements:

---

[14] The question may be raised as to whether the IoT is an infrastructure in its own or just numerous Private Area Networks (PAN) using mainly the existing DNS infrastructure on top of a TLD.

Table 3.4. Scope for IoT Governance [14].

Major Fields of Governance	Decisions and Instruments
Governance of Technical Resources	**Industrial Policy and Innovation**
Spectrum Management	International Agreements
Standardisation	Co-production
Interoperability	Upstream and Downstream Consultation
Governance of Electronic Network Architecture	**Internationalisation**
Naming and Addressing	Distributed Centralisation
Evolution or Revolution	End-to-end principle (neutrality)
Socio-political Governance	**Building Trust**
Regulations	Directives, Laws, Contracts
Self-regulation	Ethical Codes, Charters, Guidelines
Ethics	Public Debates, Education, Cultural Diversity

- Guarantee the uniqueness of identifiers which are linked to objects
  — IPv6 address or RFID chips
- Guarantee the security and stability of the network(s) which link objects (critical information infrastructure protection dimension)
  — Encryption, key registration, verification, authentication, QoE, QoS...
- Avoid monopolisation of data control and support competition among service providers
  — Multiple roots, decentralised and federated management, transparency, accountability...
- Avoid the misuse of data which emerge as a result of communication between individuals and objects
  — Privacy-by-design, PETs, privacy-by-default, "rendez-vous"[15] issues, right to the silence of the chips...

Over the next few years, the "triple play" for collaborative research and development (FP7), pan-European pilot projects (ICT Policy Support Programme), and (self-)regulation should provide public authorities with the

---

[15] The "rendez-vous" is the moment when the "object" (with the RFID or IPv6 address) meets the "subject" (data subject) at the selling point.

requirements and options to developing a clear, comprehensive, balanced, efficient and effective approach to IoT governance.

3.3.5 The Role of Standards

Over the past few years there have been a high number of proprietary or semi-closed solutions to the IoT. Along with the "My application is specific" syndrome, many non-interoperable "solutions," based on different architectures and protocols, have emerged to address specific problems. Consequently, deployments of IoT applications have been limited in scale and in scope, actually limiting the IoT to the dimension of a set of "Intranets of things." With the emergence of a myriad of applications for interconnecting billions of objects that can sense things like power quality, tyre pressure, and temperature, and that can actuate engines and lights, the need for IoT standards seems to be obvious and a matter of urgency. However, the situation is complex for at least three reasons:

- Standards for IoT[16] are based on a legacy of norms that are at the core of the Internet (DNS, TCP/IP, etc.) and the barcode system, which naturally constitutes a factor of limitation of certain developments. In the background looms the current international debate on the future of the Internet: *Evolution* (from IPv4 to IPv6) or *revolution* (clean slate approach to Future Internet design)?
- Today, the standardisation of networks is performed by private industrial groups (EPCglobal, AIM Global, ZigBee Alliance, IPSO, etc.) or by international and regional organisations (ISO/IEC, ITU, CEN, ETSI, etc.). However, the potential impact of IoT deployment on society is so great that a consensus between private actors and public authorities with respect to the choice of standards looks like a worthwhile option for ensuring that sole economic interests will not eventually prevail over public policy considerations. In this respect, the RFID Mandate (M/436) represents a relevant and useful model [15].

---

[16] There is much work going on regarding IoT standards, especially RFID, sensor networks (ISO/IEC JTC1 and IEEE802), NFC (IEEE802.15.4, Wi-Fi, etc.), M2M (M2M/MTC in 3GPP).

- To avoid fragmentation of standards, which would hamper the development of innovative applications and services, it is essential to promote granularity of standards (i.e., ensuring that each standard is sustainable on the international, regional and national level and at the level of each specific industry) as well as interoperability between different standards and, more important, different implementations of standards. The aggregation approach followed by ETSI, i.e., a system of systems acknowledging the legacy of standardisation and trying to identify and implement common functional components for the various use cases (Transport, Mobility & Logistics, Health, Energy Efficiency, Smart Metering, etc.), also deserves high consideration.

In the context of the CASAGRAS2 horizontal action, ETSI carried out in 2010 a survey of countries to know which their three first standardisation challenges of IoT were. The results are given in the table below, in which the emphasis on privacy and security is highlighted. What is most striking is the commonalty of perspectives and priorities in all the countries surveyed.

3.3.6 Law or Self Regulation? The RFID PIA Model

Making recommendations for the possible creation of an adequate legal and policy framework for the IoT is a key priority of the Expert Group on the Internet of Things. Beyond the theoretical approaches to regulation vs. self-regulation [16, 17] and the intellectual differences between disciplines, industries and continents, it seems obvious that technology evolves faster than legislation.[17] By making national parliaments a new actor in the EU legislative-making process through the mechanism of subsidiarity control, the Treaty of Lisbon has resulted in further complexity in the development of EU law (directives and regulations). The complexity of the decision-making process can be seen on the flow chart indicated in [18].

In this context, continued self-regulation, for example in the form of voluntary codes of conduct, seems to be the most effective means of protecting privacy while fostering innovation. This is particularly true in the EU where

---

[17]*"Unfortunately, the legislation tends to lag behind technological advancements,"* interview with ENISA Risk Management Expert, Barbara Daskala, 4 May 2010.

Table 3.5. Key IoT standardisation challenges.

	Priority 1	Priority 2	Priority 3
Australia	Identity Architecture	Communication Networking	Security Privacy
Brazil	Numbering Naming/Identification Governance	Integration Scalability Resilience Autonomic Systems Trust	Communication Technology Network Technology
China	Numbering Naming/ID Governance	Communication Technology Network Technology	Security Privacy
Europe	Architecture for Interoperability Resolving Techniques for Existing Numbering, Naming/Identification Systems & Privacy Integrity	Standards Meeting Evolving Internet and IoT Governance Requirements Integrated Discovery Services Service Federation, Scalability	Privacy-by-Design Security-by-Design Health-by-Design In-network Data Management Resilience Safety
India	Architecture Context Aware Management	Security Privacy	Energy Efficiency Sustainable Communication and Networking
Japan	Numbering Naming/Identification	Governance Interaction between Autonomous AAA Systems	Security Privacy Context Awareness and Service Discovery Interaction between Systems
Korea	IoT Architecture	IoT Identification and Resolution	IoT Management
Malaysia	Numbering Naming/Identification	Communication Networking and Discovery	Security Privacy
Russia	Architecture Numbering Naming/Identification	Communication Networking Integration	Security Privacy
USA	Numbering Naming/Identification Governance	Security Privacy	Communication Technology Network Technology

(*Source*: ETSI, with the courtesy of Patrick Guillemin).

there is already a clear, powerful and effective legislative framework, including the Data Protection Directive (1995) and the ePrivacy Directive (2002). Further strict regulation for imposing privacy obligations on industry in the use of specific ICT could stifle innovation as it would be outpaced by technological developments and user preferences.

The endorsement by the Article 29 Data Protection Working Party of an Industry Proposal for a Privacy and Data Protection Impact Assessment (PIA) Framework for RFID Applications, on 11 February 2011, constitutes an interesting model that could be used for other situations or areas, such as

Smart Metering and Online Behavioural Advertising. Such a model is situated somewhere between self-regulation and co-regulation.

On one hand, self-regulation is *"the possibility for economic operators, the social partners, non-governmental organisations or associations to adopt amongst themselves and for themselves common guidelines at European level (particularly codes of practice or sectoral agreements)."*[18] Self-regulation is therefore entirely composed of voluntary initiatives and does not require the adoption of an empowering and preliminary legislative act. Contrary to the system of voluntarism, self-regulation involves a group of private actors that regulates the conduct of its members on the basis of certain norms. These norms are eventually compiled in the form of a code of conduct.

On the other hand, co-regulation — at least as it is defined in Europe — refers to the situation in which a Community legislative act entrusts the attainment of pre-defined objectives to parties that are recognised in the field. The context of co-regulation is therefore, and voluntarily, set by a legislative act. *"Co-regulation combines binding legislative and regulatory action with actions taken by the actors most concerned, drawing on their practical expertise. The result is wider ownership of the policies in question by involving those most affected by implementing rules in their preparation and enforcement."*[19] The interaction between different instruments of regulation makes here the regulatory intervention of many private and public actors possible. This interaction links indeed an instrument of self-regulation – a set of guidelines, a code of conduct, etc. — and an instrument of regulation, and, thus, acts as a restraint that may take the form of administrative or legal sanctions, should the need arise.

Point 4 of the RFID Recommendation says: *"Member States should ensure that industry, in collaboration with relevant civil society stakeholders, develops a framework for privacy and data protection impact assessments. This framework should be submitted for endorsement to the Article 29 Data Protection Working Party within 12 months from the publication of this Recommendation in the Official Journal of the European Union."* The objective stated here (i.e., to develop a framework) was obviously hard to attain because the actors ("Member States," "Industry," "relevant Civil Society stakeholders") were vaguely defined, had conflicting interests difficult to reconcile,

---

[18] Interinstitutional agreement on better law-making, 2003/C 321/01.
[19] COM(2001) 726 final.

especially within a short timeframe, and were supposed to produce a result that would still need to be endorsed by the European Data Protection Authorities. The self-regulatory system could not reliably succeed because of the antagonism between the different parties and the lack of a driving force or mechanism. The co-regulatory system was not relevant because there was not a Community legislative act per se — only a recommendation. Therefore, the route that the EC opted to follow was a hybrid of self-regulation and co-regulation, with all the stakeholders, except the Article 29 Data Protection Working Party, interacting with themselves and the EC straining itself to transcend interest-based conflicts and "turf battles." Finally, it took Industry 10 months to work out and submit its initial proposal (from May 2009 to March 2010) and another 10 months (from April 2010 to January 2011) to make it a final proposal matching the requirements from all the actors, including the EU national DPAs. The Article 29 Data Protection Working Party was all along the process a discrete but very active actor to such good purpose that the formal endorsement occurred only one month after the submission of the final proposal.

The Table below gives the key milestones for this successful endeavour.

The example of RFID is indeed a good model to analyse and replicate. Yet, even for technologies that are developing fast like the Internet of Things, the advantages of a system of self- or co-regulation (e.g., institutional knowledge, flexibility, lower cost, better timing) are not automatic; they can only be obtained if a number of elements are brought together:

- Cooperation: all members of the community of stakeholders must be committed to cooperating with each other in order to implement the system and address complaints;
- Effectiveness: the system must have a timely response rate, be flexible and applied in both the spirit and the letter; it must also be regularly reassessed;
- Efficiency: the community of stakeholders must be able to handle individual complaints, without charge, and foresee adequate and credible sanctions to support its decisions;
- Resources: the system must be sufficiently resourced;
- Compliance: the system must fully comply with the law (i.e., the Data Protection Directive) and never deprive a citizen of the protection by the law;

Table 3.6. Key milestones of the RFID PIA framework.

Date	Stakeholder	Milestone	Remarks
12/05/2009	EC	Adoption of the RFID Recommendation	Ref. C(2009) 3200.
17/06/2009	A29WP	Plenary meeting	EC presents the RFID Recommendation and announces plan to create an "Informal Workgroup" on its implementation.
08/07/2009 to 28/04/2010	EC, Informal Workgroup	5 meetings	Focus on PIA Framework.
26/01/2010	EC	Mail to A29WP Chairman	EC submits to A29WP an informal draft Industry Proposal for an RFID PIA Framework.
02/03/2010	A29WP TS	Preliminary review of informal draft Industry Proposal	Informal feedback sent to Industry.
31/03/2010	Industry	Formal submission of Proposal I to EC & A29WP	
16/06/2010	A29WP TS	Meeting on Industry Proposal I	ENISA attends and provides comments.
27/06/2010	Informal Workgroup Rapporteur	Delivers final report of the Informal Workgroup on the RFID PIA Framework	
08/07/2010	AIM Germany and BSI	Publication of Technical Guidelines RFID as Templates for the PIA-Framework	
13/07/2010	A29WP	Opinion 5/2010 on Industry Proposal I	No endorsement: Risk assessment Tags carried by persons RFID in retail sector Level definitions Stakeholder consultation phase Special categories of data Security and privacy by design
14/07/2010	ENISA	Agency Opinion on Industry Proposal I	In line with A29WP Opinion.
19/07/2010	A29WP Chairman	Informs Industry of A29WP position on Proposal I	Letter to GS1 Global Office, AIM, EABC, ERRT.

(*Continued*)

Table 3.6. (*Continued*)

Date	Stakeholder	Milestone	Remarks
06/10/2010	BITKOM	On behalf of the German industry (BITKOM, AIM, specific industry sectors), BITKOM officially informs the EC of "an enhanced version of the RFID PIA Framework" (Proposal II)	Key concerns about Proposal I: Not enough related to EU legislation the definition of "privacy levels" is too vague a "small scale" PIA is not foreseen no clear guidance on what companies need to do for a PIA mitigation strategies are unclear.
21/10/2010	German industry bodies	Release of Industry Proposal II (alternative Proposal)	
22/10/2010	EC, ENISA, Industry	Meeting in Frankfort to reconcile the two "competing" Industry Proposals	Meeting chaired by EC, moderated by ENISA. Next steps: an Industry Proposal III (I + II) will be prepared by (1) combining the "decision tree approach" of Proposal II with the "privacy level" approach of Proposal I and (2) including the risk-assessment-methodology of Proposal II.
04/11/2010	Industry	Submission of Proposal III to A29WP	
16/11/2010	A29WP TS	Meeting	Corrections are requested on Industry Proposal III.
30/11/2010	A29WP TS	Informal feedback on Industry Proposal III	Sent to EC & Industry after internal consultation process on the draft TS report. Meanwhile, over the period 17 to 25/11/2010, the EC stimulated Industry to reach consensus on the likely A29WP comments.
03/12/2010	Industry	Industry submits its Proposal IV to A29WP	

(*Continued*)

Table 3.6. (*Continued*)

Date	Stakeholder	Milestone	Remarks
21/12/2010	A29WP TS	Corrections on Industry Proposal IV	TS sends EC & Industry some editing amendments to Industry Proposal IV.
11-12/01/2011	Industry	Differing views within Industry on text about RFID in retail sector	As discussions within Industry have reached a stalemate, EC informally contacts A29WP TS and offers to Industry a solution which is accepted by all industry bodies.
12/01/2011	EC	Final Industry Proposal V is uploaded onto CIRCA and forwarded to A29WP	A29WP now prepares a draft Opinion.
27/01/2010	A29WP TS	Draft Opinion is to endorse Industry Proposal V	A written procedure is considered to save time, but it will not be used because of the proximity of the date of the next A29WP plenary meeting.
11/02/2011	A29WP	Opinion 9/2011 endorses RFID PIA Framework	Unanimous agreement among DPAs.

(*Source*: Gérald Santucci).

- Independence: the decisions taken by the community of stakeholders must be made independently of specific interests and interest groups;
- Transparency and accessibility: citizens (e.g., consumers, patients, users of public transport) must have an easy access to their data which is handled by the self- or co-regulatory system.

3.4 Outlook

IoT together with the other emerging Internet developments, including the Future Internet, are the backbone of the digital economy and the digital society. There are reasons to believe that a time when devices are getting smaller than the period at the end of this sentence is poised to be a time of opportunity; but there are also reasons to believe that it might be a time of risk.

The IoT is just emerging and its deployment will probably unfold gradually over at least two decades. Within the timeline that IoT developments are likely to occur, it will be important for Europe to monitor various signposts pointing out the direction and pace with which the field is advancing and to assess any resulting potential opportunities and risks for its strategic and socio-economic interests. Such signposts should include, in particular, the size and nature of demand, the speed of dissemination of IoT technologies into vertical application areas, advances in adaptive software, advances in IoT underlying technologies such as miniaturisation and energy-efficient electronics, and efficient use of spectrum.

Getting bird's eye view of the evolution of the IoT, we can identify four "grand challenges" which Europe should make as its duty to take up:

- To keep at the forefront of technological innovation and research capability. Breathtaking advances in ICT and in the convergence between ICT and Nanotechnology are taking place at exceptional speed. Europe must be abreast of these developments;
- to make that progress in IoT will allow addressing the societal challenges of our times. The IoT lies at the confluence of the various technologies that can tackle the challenges of climate change, energy efficiency, mobility, health, and so forth. Innovation in IoT — new products, services, interfaces and applications — is therefore essential to create smart environments and smart spaces. As science and technology is the spark of human progress, we must overcome the limits to our vision and work each day to make progress a reality in the lives of people;
- to see that the benefits of the IoT revolution are shared by all citizens of the world, not just those fortunate enough to live in Europe and other developed nations. Europe must lead in compassion as it leads in innovation. The greatness of our hearts must match the greatness of our inventions;
- to create a "new alliance" of government and business to hit the right balance between the potential socio-economic opportunities and threats of the IoT. Such a new alliance is likely to involve a combination of strict regulation, self-regulation and co-regulation — but the direction along which the pendulum will swing remains unclear today.

If we begin to make tangible progress in some key policy areas — privacy, security, ethics, inclusion, etc. — we will be off to a promising start in this new decade. We are living in dramatically new times in which the best coasts along the worst. Research in IoT has the ability to surprise and enthral and frighten as well. It's a matter of perspective, opinion and perception. But in many respects, we all have enduring dreams as old as our nations — health, mobility, safety and security, success in life, etc. IoT provides one of the solutions to meet our best hopes and ideals and to master the awesome challenges we face today. It offers unprecedented opportunities to vastly improve the health of the planet as well as the health and well-being of individuals. We must not let this opportunity slip by.

Acknowledgments

Writing this paper would not have been possible without the "invisible army" of helping hands that unknowingly dedicated their knowledge to make this initiative a hopeful move. Among the countless people who inspired me for this work, I would like to give a special thanks to all the individual members of the former RFID Expert Group (2007–2009) and current Internet of Things Expert Group (2010–2012) who will recognise themselves in the paper. I want to take this opportunity to thank Ovidiu Vermesan for his outstanding conduct of the IERC and for making this cluster-book possible. Finally, I thank all the colleagues from my unit for their dedication and effectiveness as well as Hana Pechackova, my colleague from DG JUST, for her ceaseless efforts in helping me keep the compass right during the journey to the RFID PIA Framework.

The opinions reported in this paper are purely the author's and do not necessarily represent the views of the European Commission. Any mistakes are the sole responsibility of the author.

References

[1] IERC — European Research Cluster on the Internet of Things. http://www.internet-of-things-research.eu/.
[2] P. Gautier, "Internet des Objets: Objets 'connectés', objets 'communicants'… ou objets 'acteurs'?," 2010.
[3] D. Hendricks, "Internet Devices Become Collaborators," http://ideasproject.com/docs/DOC-532.
[4] New York Times, This data isn't dull. It improves lives, 12 March 2011.

[5] G. Santucci, "The internet of things: Between the revolution of the internet and the metamorphosis of objects," *Vision and Challenges for Realising the Internet of Things*, CERP-IoT (now IERC), March 2010.

[6] J. Attali, "Une brève histoire de l'avenir," Fayard, 2006, Chapter 4, in particular, pp. 248–269.

[7] Commission of the European Communities, *"Internet of Things — An Action Plan for Europe,"* COM (2009) 278 final, Brussels, 2009, http://ec.europa.eu/information_society/policy/rfid/documents/commiot2009.pdf.

[8] Commission Decision of 10 August 2010 setting up the Expert Group on the Internet of Things, 2010/C 217/08, http://eur-lex.europa.eu/LexUriServ/LexUriServ.do?uri=OJ:C:2010:217:0010:0011:EN:PDF.

[9] EUR-Lex — Access to European Union law, "Directive 95/46/EC of the European Parliament and of the Council of 24 October 1995 on the protection of individuals with regard to the processing of personal data and on the free movement of such data", 31995L0046, http://eur-lex.europa.eu/LexUriServ/LexUriServ.do?uri=CELEX:31995L0046:EN:NOT.

[10] European Commission, "A comprehensive approach on personal data protection in the European Union," *COM(2010) 609 final*, Brussels, 2010, http://ec.europa.eu/justice/news/consulting_public/0006/com_2010_609_en.pdf.

[11] V. Mayer-Schonberger, "The Virtue of Forgetting in the Digital Age," Princeton University Press, September 14, 2009.

[12] ENISA, "Flying 2.0 — Enabling automated air travel by identifying and addressing the challenges of IoT and RFID technology," 12 April 2010.

[13] The New York Times, "The Data-Driven Life," 26 April 2010.

[14] P.-J. Benghozi, S. Bureau, and F. Massit-Folléa, "The internet of things: What challenges for Europe?," *Editions de la Maison des Sciences de l'Homme*, juillet 2009, p. 142.

[15] CEN — European Committee for Standardization, "Radio Frequency Identification (RFID)", 2009, http://www.cen.eu/cen/Sectors/Sectors/ISSS/Activity/Pages/RFID.aspx.

[16] R. H. Weber and R. Weber, "Internet of Things: Legal Perspectives," Springer, June 2010.

[17] R. H. Weber, "Internet of things — New security and privacy challenges," *Computer Law & Security Review*, 26(1), pp. 23–30, January 2010.

[18] European Commission, "Codecision Flow Chart," 30.07.2010, http://ec.europa.eu/codecision/stepbystep/diagram_en.htm.

4

Between Big Brother, Matrix and Wall Street: The Challenges of a Sustainable Roadmap for the Internet of Things

Dr. Alessandro Bassi

IoT-A Project, France

4.1 Introduction

The path to success for a fully-fledged implementation of intercommunicating objects has quite a few roadblocks, belonging to different areas, such as sociology, technology, and business.

The original, main goal of the precursor of the Internet (the ARPAnet) was resilience. Back 40 years ago, architectural choices were made in order to provide the highest robustness, as the network was designed to sustain a nuclear strike: the centrality of the IP protocol, a connectionless, unreliable network protocol, and packet-switching at hop points provided unprecedented robustness, allowing the original ARPAnet to grow up to the current size. The military support for the TCP/IP protocol suite helped to secure the development of the original network. Later on, universities were allowed to take on the development of the network, following the original design of connecting network nodes and not having a fixed path between two endpoints.

However, as the complexity of Internet grew, due to a large increase in the heterogeneity of connected devices and services, novel architectural models were developed and applied: Overlay networks and autonomic management are just two main examples of solutions implemented in order to cope with the size and diversity of today's internet.

With the emergence of the Internet of Things (IoT), though, all management and orchestration mechanisms developed for today's networks are highly insufficient in dealing with the vast amounts of information that small connected objects are bound to generate when they become fully deployed.

Even though current Internet protocols are mostly proven to be scalable and that the Internet community was rapidly able to provide solutions for any problem that may emerge, many characteristics of the IoT are fundamentally different, making the current Internet model inadequate. For example resource constrains in IoT is reversing the phenomenon of software induced hardware obsolescence. To date if software grows we just add hardware; within the IoT boundaries, however, there is no possibility of over-provisioning hardware; and software has to follow, which means that standard approaches to service design are bound to fail. Another example is the area of security and privacy due to the trend that hundreds or thousands of information objects will be tightly integrated into all aspects of every single human. Access control and management to information is an overwhelming endeavour for any individual, so that totally new approaches are required.

While there's no single definition of IoT, we can assume that every object that is identifiable, is able to sense the environment around it, and is able to modify the environment, belongs to the IoT. To this end, IoT comprises a digital overlay of information over a highly heterogeneous physical world of objects. In the coming future, such objects are expected to outnumber the human population by a factor of three. Therefore, it is paramount to improve the best-known techniques for managing and orchestrating heterogeneous devices, and develop new models and visions, that can cope with the complexity and the scalability issues that the IoT will bring.

4.2 Sociological Issues

Currently, IoT technologies have a "Big Brother"-like image, mainly coming for the usage of RFID identification means.

The acceptance of IoT is strongly linked with basic privacy and personal data respect. Today, technologies that allow the identification of objects, such as RFID, can be seen as violating the privacy regulations of the European Union. There are literally hundreds of examples, from health monitoring systems processing inhabitants' sensitive data to payment systems to commercial spam, that show how novel technologies could be misused.

Therefore, a prerequisite for trust and acceptance of these systems is that appropriate data protection measures are put in place against possible misuse and other personal data related risks.

For some experts, efforts in this sense are superfluous, as they argue that the uptake of IoT will change our perception of privacy, citing as evidence the amount of information that people, and especially the younger generations, are willing to display publicly on social networks, and the global uptake of technologies such as mobile phones and credit cards, able to track any of our habits, at any time. However, while the above argument is not unreasonable, it is common opinion that is the need to develop novel information security mechanisms, if for anything, just in case social trends or legislation change. As a reminder, the TCP/IP protocol suite was developed in the early 70, and we can easily affirm that the original team was not imagining the current deployment of the protocol suite they were designing at that time.

As in any other field, applying privacy and security as a patch after the design phase leads invariably to less-than-satisfactory results and cumbersome architectures. As IoT can still be considered in his infancy, developing architectures with those clear goals in mind is not only good practice, but also will likely save an enormous amount of effort afterwards.

Regarding security issues, the main difference between the IoT and the Internet is in the computational capabilities and in the easiness of possible attacks. While the Internet is somehow a closed system, composed of devices having rather large computation and storage facilities, allowing complex firewall and secure architecture at entry-points, an ubiquitous IoT will have billions of possible entry points to secure, Moreover, as the devices normally have very limited computational and storage power, there is limited possibilities of running complex software.

4.3 Technological Challenges

Given the complexity of the IoT paradigm, involving the coexistence of many types of devices, communications and networking technologies, applications, and the list of technological challenges to be addressed is potentially very long.

4.3.1 Energy Management

Energy in all its phases (harvesting, conservation and consumption) is a major issue, not only in the IoT area, but more in general for the society at large. The development of novel solutions that maximize energy efficiency is paramount. In this respect, current technology development is inadequate, and existing processing power and energy capacity is too low to cope with future needs.

The development of new and more efficient and compact energy storage sources such as batteries, fuel cells, and printed/polymer batteries, together with new energy generation devices coupling energy transmission methods or energy harvesting using energy conversion, as well as extremely low-power circuitry and energy efficient architectures and protocol suites, will be the key factors for the roll out of autonomous wireless smart systems.

Research efforts will focus on multimodal identifiable sensing systems enabling complex applications such as implants monitoring vital signs inside the body and drug delivery using RFID, whilst harvesting energy from different sources.

4.3.2 Communication Fabric

The existence of several mechanisms and protocol suites developed to transfer information between peers within what can be indicted at large as intercommunicating objects area is similar, in many ways, to the beginning of the Internet. More than twenty years ago, different technologies were developed to transfer data efficiently between somehow homogeneous computing machines; most of those efforts have been abandoned, and the TCP/IP protocol suite emerged as the main information carrier for the Web.

Objects need to speak a common language, if the Internet of Things wants to provide a useful fabric for the development of services and applications. Therefore, the IoT must find a "narrow waist," similar to the IP protocol for the Internet, in order to provide a seamless communication flow between entities, intended as devices or services. While it is arguable where, in the protocol stack, this narrow waist will be, which protocols will belong to it, and through which channels devices will be able to interoperate, full IoT development will not happen without a single, uniform, and universal view.

The different communication protocols and mechanisms, such as Bluetooth, Zigbee, WiFi, WiMAX, or NFC were all developed with a specific target in mind. These protocols follow a different concept: while Internet is the network of networks, they are more suited to a so-called "Intranet of Things," where solutions are applicable for a single problem, missing the generality that distinguish the TCP/IP protocol suite.

The creation of a generic Yet-Another-Wireless-Protocol is probably doomed to fail; instead, as TCP/IP was not created by scratch, an evolution of current protocols, with the extension of some protocols, the creation of gateways between different stacks, and the development of cross-protocol routing elements will provide the necessary uniform communication fabric that guarantees device interoperability.

4.3.3 Beyond IoT-A: Scaling Up the IoT

The Internet of Things Architecture (IoT-A) project aims at developing a reference architectural model able to provide a uniform view or the future IoT. IoT-A covers technological issues varying from low level hardware issues such as common cryptographic algorithms, protocols and communication mechanisms, addressing and naming schemes, up to high level protocol integration issues such as machine-to-machine (M2M) interfaces and plug-and-play objects recognition and configuration, while aspects related to the integration of IoT in the generalized context of user services are beyond scope. To this end, there is a need of complementing the IoT-A architecture by focusing on the definition of a uniform service interface to enable generic integration of the interconnected object functions in the context of user services, taking special care of preserving their autonomous operation capability.

In addressing these requirements it seems necessary to adopt the autonomic communications paradigm that allows self-* properties to be designed, implemented and deployed thus creating a thin, modular, and autonomic middleware layer in-between the user applications layer and the underlying IoT-A network architecture, wherein autonomic algorithms can be readily deployed as generic means for service implementation and objects management. Such autonomic properties can be either generic, taking care of common management and orchestration object operations or context-specific driven from several application domains.

In addition, the vast volumes of generated information give rise to an information service as constituent of the autonomic layer that conceptually interconnects objects and IoT applications. Such a service must be capable of modelling, identifying, searching, retrieving, aggregating and delivering information requested by any entity (application or object) in a way that is compatible with the entity's interface, thus advancing interoperability in IoT.

This development is also a necessity for business development. As orchestration and management mechanisms can shape future business models, and create new actors and modify greatly the value chain, a special attention needs to be put on those issues.

4.3.4 From Objects to Smart Things: Integrating Capabilities into Materials

The integration of chips and antennas into non-standard substrates like textiles and paper, or even metal laminates and new substrates with conducting paths and bonding materials adapted to harsh environments and for environmentally friendly disposal, will become a mainstream technology. RFID inlays with a strap coupling structure will be used to connect the integrated circuit chip and antenna in order to produce a variety of shapes and sizes of labels, instead of direct mounting. Inductive or capacitive coupling of specifically designed strap-like antennas will avoid galvanic interconnection and thus increase reliability and allow even faster production processes. The target must be to physically integrate the RFID structure with the material of the object to be identified, in such a way as to enable the object to physically act as the antenna. This will require ultra-thin structures ($<10\,\mu$m), as well as printed electronics, which are both robust and flexible.

4.4 Business Considerations

The advent of Internet had the effect of an earthquake on "old" business models, some of them established centuries before. Similarly, a full realisation of the IoT paradigm is likely to recreate the same effect, revolutionising the current business models and value chains and drastically changing main actors.

According to some estimates, IoT could bring added value services up to 200 billion USD/year. However, today, despite technological advances that could realise a much higher penetration, the current impact is still tiny. Small pilots have been funded, but normally the incompatibility of different experiences and technologies makes them ineffective to reach the necessary level of capital investments to create the snowball effect typical of technological revolutions.

Currently, IoT is seen almost uniquely as service enabler, similar to the smart phone market, where the "killer application" came from the ecosystems of applets available for commercially successful platforms such as the iPhone, Blackberry and Android. However, IoT has the power of changing the inner structure of business: Not only creating new high-end services over known platforms but allowing radically new business models. In the same mobile world used as a base for IoT business models, the very emergence of mobile communication and the possibility of using underlying technologies in a different way than originally thought. Voice-over-IP is a clear example of this: while packet-switching emerged as a main data carrier mechanism, in contrast with typical circuit-switching technologies used for voice, it is now used to carry "voice packets" at a fraction of the price, if charging at all, that old operators used to charge for international communications. This had a huge impact on business models of traditional telecom operators, and forced them to change their business model to survive.

Moreover, the possibility of uniquely identifying any object will allow easier renting and co-owning of many different objects. A typical example can be a drill: a device that is bought for 100 Euro or more, and used very rarely. Considering the average cost-per-use, the price of each hole ever made by the drill is huge — without considering the space occupied in the cellar. Therefore, many objects we currently own but use very seldom can be co-owned or rented, creating new services and new actors, if they could be uniquely identified and their usefulness verified at any time. Identification technologies such as RFID, and sensors could effectively realise true pay-per-use schemes, which can also be successful because of a different social attitude that can be seen from the success of Facebook and other social networks.

4.5 Future Works

The Internet of Things needs effort in many directions in order to achieve a full implementation. First of all, technical issues need to be convincingly addressed, and efforts in this sense have already been made by the European Commission in recent calls of the Framework Programme 7. Technology advances will then enable new business models, that have to be developed together with convincing adoption plans. Last but not least, governance issues must be solved at National and European level, developing the sense in EU citizens that the IoT revolution will not bring any threat, but significant advantages to the population. All these advances will not be possible without a clear engagement and common roadmaps from the private and the public sector for the foreseeable future.

5
2011: A Decisive Year

Rob van Kranenburg

Council, a Think tank for Internet of Things, The Netherlands

5.1 Introduction

Technicism and science are consubstantial, and science no longer exists when it ceases to interest for itself alone, and it cannot so interest unless men continue to feel enthusiasm for the general principles of culture. — José Ortega y Gasset, The Revolt of the Masses.

In this short text I want to propose two things for IoT: an open generic value, service and organizational layer and a device. Combined, these can create enthusiasm and a vision to transcend the false opposition between "progressive" Wikileaks and "conservative" US State Department calls for the development and support of web-based circumvention technology to enable users in closed societies to get around firewalls and filters in acutely hostile Internet environments [1]. In terms of Climate Change, running out of natural resources and exploding human birth rates, defining what is an open or closed society in human terms of flows of data is an academic exercise when more cities start to look like New Orleans, Detroit and Brisbane.

5.2 Impact

Co-moderating the first IOT-A Stakeholder workshop in Paris (October 2010) it became clear how strongly big industry and the big system integrators envisage a global generic value chain where items will be logged, tracked and traced.

The extent to which this will change them as individual brands is moderately foreseen in this operation. In the Spring 2011 edition of the Situated Technologies series, Christian Nold and I argue in 'The Internet of People for a Post-Oil World' that this process much resembles the standard-making of the barcode and EPC Global. The first refers to items as a batch, the second (RFID) to uniquely identifiable items. As the Internet of Things has huge societal as well as economic implications, we argue that the standard making process that will build this generic value chain should include social and cultural issues as well as logistic and system driven ones.

The second reason why we need global hard regulated standards on a simple, effective and highly integrated value chain wave strategy with as little noise, redundancy and overlap as possible is that we have as of yet no coherent academic view on the effects of radio-waves on humans, animals and plants. The Sensing Planet, and Smart City are very powerful concepts to streamline new balances between people and the planet, people and animals and plants and people and other people. They require huge sensor-beds and a large number of UHF readers as well as astronomical amounts of handheld readers in mobile devices. In the input I gave to EP Lena Kolarska Bobinska a number of the issues I raised through Council, the thinktank I founded in december 2009 (currently 71 professionals), made it to Parliament resolution of 15 June 2010 on the Internet of Things (2009/2224(INI), (P7_TA-PROV(2010)0207) such as "Stresses the importance of studying the social, ethical and cultural implications of the Internet of Things, in the light of the potentially far-reaching transformation of civilisation that will be brought about by these technologies; takes the view, therefore, that it is important for socio-economic research and political debate on the Internet of Things to go hand in hand with technological research and its advancement," and the request for the study of "the impact of electromagnetic fields on animals, especially birds in cities." In the light of the recent mass deaths of birds in different places in the world linked by some to "microwave radiation from 4G-networks" [2] this seems to become relevant. It makes sense to have a Sensing Planet only if the very infrastructure that does the sensing becomes not an integral part of the problem.

Without significant changes in the next 5 years of EU's Digital Agenda the billions invested so far and in FP8 will have served as investments in platforms that are either closed (FP7 and 8 Security), used for free by an integrated Chinese value layer (that has won from GS1 as the main dominant identification scheme) serve as a municipal layer of citizen services dominated by the IBM/Cisco Smart City concept, and facilitate a service layer dominated by Facebook/Google/Apple as the social networking billions will find their way into more critical functions of everyday life. Europe will be a testbed for these services with its high level of education of citizens and rich culture that can be explored. But it is clear that it will pay all the bills and make no money.

In the mid-1990's we see the first attempts to give citizens real-time feedback on public transport in cities. The European Union had a very productive R&D scheme for academics and companies called "Intelligent Information Interfaces" which included projects like "Ambient Agoras," "Interliving" and "Grocer." In 1996, Philips developed a social RFID project called Living Memory (LiMe) that would "provide members of a local community with a means to capture, share and explore their collective memories" via RFID tokens that interact with screens in bars and bus stop information boards. Hardly any of these projects led to the creation of actual products. Companies could not agree on the intellectual property for the things that were developed and the timing for pervasive computing business models was off. They did not produce anything but they have learned from these projects. Today they talk about co-creation and real people. In 2003, at a time when the EU funded research projects like LiMe, Steven Kyffin of Philips stated:

USEFUL... listening to and developing technology for ordinary people sums up what we might refer to as Co-Creative design. Involving the end user in a core and proactive manner at all stages in the product or system creation process.

RELEVANT... listening to and developing technology for ordinary people is so relevant because the "ordinary... ness" is the issue.

Malta has been a member of the European Union since 2004. Like all states today it aims to be smart. It has set up Smart City Malta. The contact officer providing the expertise is from Dubai, the press contact from Cisco. The model on which it is based is... Dubai: "SmartCity Malta is a self-sustained industry township for knowledge-based companies located in the Ricasoli Estate in Malta. To be developed by SmartCity, in partnership with the Government of Malta, SmartCity Malta will be home to a vibrant knowledge-economy community anchored by leading global, regional and local companies. SmartCity Malta is set to become the leading ICT and Media cluster in the heart of the Mediterranean and the first European outpost of the global SmartCity network.

With a minimum investment outlay of US $300 million, SmartCity Malta will transform Malta into a state-of-the-art ICT and Media business community based on the successful clusters of Dubai Internet City, Dubai Media City and Dubai Knowledge Village" [3].

This is not serious. European citizens are getting the worst of all possible worlds. They do not have the positive effects of the huge potential and actuality of policy, product, and hardware infrastructure integration of the Chinese vision, nor the positive effects of the USA extreme no-policy-whatsoever laissez fare situation that allows the social networking sites to flourish without privacy policies, Apple to close garden and Google, IBM and CISCO to start investing in society critical infrastructure and services in purely capitalist fashion. Instead their money pays for platforms that other people make money off, their ISP's have to comply with data privacy rules based on 19th century notions of romantic autonomous individuals, there is no VC money for European IoT startups that have to go to Asia to get funding, and worst of all: We all know this and seem to simply carry on, caught in our policy formats.

Uptake goes through devices. People run their applications on their smartphones without investigating the algorythms that inform them. Code is law in this respect still. The device holds the key to the protocols it allows on the back end and the applications and interoperability it allows with other hardware and software. It is possible to embed social, cultural and economic value in this process. In fact that is what law is doing all the time.

Speaking on The Internet of Things at a Far Horizon workshop held on the 2nd and 3rd of December 2010 on "Education for an ICT revolutionizing society" that was held within the Seventh Framework Foresight workprogram in the EU I was engaged in conversation with Constantijn van Oranje-Nassau who in his introduction had spoken of the heavy rucksacks his children were taking to school. He wondered when the books could go in favour of a small computing pad. Someone uttered: Let's buy them ipads! In the following discussions it was suggested to develop European standards for educational tablets, rather than adapting education to existing concepts and standards that have been developed by some non European major players. Recently a Singapore a pilot was run with schoolchildren reducing the weight of their bags drastically by giving them ipads. Many schools in the US and in European countries are following. Victor van Rij, project leader of Far Horizon, mentions a Scottish school that in september 2010 fully changed to the ipad and points to the developments in the US: "The New York City public schools have ordered more than 2,000 iPads, for $1.3 million... More than 200 Chicago public schools applied for 23 district-financed iPad grants totaling $450,000. The Virginia Department of Education is overseeing a $150,000 iPad initiative that has replaced history and Advanced Placement biology textbooks at 11 schools" [4].

I fear that very soon someone in Germany or the Netherlands will propose a similar idea. Apart from the fact that this would give Apple not only 30% on any educational EU app, but also some intrinsic editorial control, I see no reason why we should spend our money making Apple shareholders richer then they are already.

EU industry could build a more robust and cheap tablet. Build in close relationship with high education potentials such as India and China it could build a device that has the potential of tens of millions of open platforms for learning. Also for learning technology and programming. Mark Belinski states [5]:

"The iPad is magic to children. Press a button and it does everything that you want it to. The problem is that it doesn't tell you how the magic happens. There's a danger in teaching kids to inherently trust technology without being more critical of it. The iPad defaults in being a consumer technology, not a producer technology.... To truly learn about how to orient ourselves to the democratic society within which we live, it's hard to believe that learning through a centralized and strict system is the best way to go. Open source software intrinsically reflects the values of our society — it is transparent, accountable, and efficient. As our government is increasingly built on this framework, from the new open data repositories to the sites themselves, as well as our corporate enterprise infrastructure. It's crucial to teach our kids with tools that reflect this world."

Uptake goes through devices. EU industry should make them. EU values should be embedded in them. EU infrastrucure should host them. A tablet for educational purposes across all member states is a very good start.

In the Council vision of a successful development of IoT there are regional "islands" connected through generic layers on mission critical services that directly address Climate Change, Peak oil and the deep social changes we are witnessing from individuals being able to organise quickly for better: More social cohesion, balance and justice, or for worse: more gated communities, selfish waste of resources and communities of like minded social media "friends." These generic layers can be steered most productively through devices. It is not inconceivable that a tablet for schoolchildren becomes a device that will serve these critical functions in a society that consists of organised networks. It can become a kind of passport. Through Near Field Communication it can serve as a bartering tool, either in money, or in "karweitjes" (small jobs). It can be linked to smart energy meters in a street or neighbourhood showing you are actually taking out energy or putting some back in. It can be the new tool to pay "taxes." Already Finance Departments are beginning to cater to their customers, showing them where their money is actually going. Where Does My Money Go? is "an independent non-partisan project trying to make government finances much easier to explore and understand — so you can see where every pound of your taxes gets spent" [6]. Open Data programs have no logical end so at one particular point taxpayers will want to see all the bills and decide if they want to pay all the costs that states and organisations claim is necessary to serve primary functions. The algorithms informing the key protocols to the device can be opened up for change every five years.This trade off between islands where developments can take place rooted in local practices (that can be even quite luddite) and a generic layer that is just, that treats resources and animals and plants as equals, can be the new iteration where activists, industry and policy can work together without compromising on their real talents.

References

[1] http://www.state.gov/g/drl/p/127829.htm
[2] http://www.buergerwelle.de:8080/helma/twoday/bwnews/stories/2041/
[3] http://www.zawya.com/story.cfm/sidZAWYA20110122085719
[4] http://www.nytimes.com/2011/01/05/education/05tablets.html?_r=3&sq=ipad&st=cse&scp=2&pagewanted=all
[5] http://www.huffingtonpost.com/mark-belinsky/horrified-by-schools-that_b_804750.html
[6] http://wheredoesmymoneygo.org/about/

6

Internet of Things — From Ubiquitous Computing to Ubiquitous Intelligence Applications

Prof. Ken Sakamura[1,2] and Chiaki Ishikawa[2]

[1] *The University of Tokyo, Japan*
[2] *YRP Ubiquitous Networking Laboratory, Japan*

6.1 Open Code Architecture: Ubiquitous ID Architecture

ubiquitous ID (uID) technology, automatically identifies physical objects and/or locations by the uIDs assigned to them in order to provide services associated with those physical objects and/or locations. uID architecture is for the realization of "Context Awareness" in Ubiquitous Computing.

Many applications have been tried using this architecture. Among the applications, we found two important areas of applications. They are application for **locations**, i.e., location information systems, and applications for **objects**: These are systems for traceability, and management of assets, etc.

In the following, we explain the principle of uID architecture in the first part, and then explain the real-world deployment at various stages (from feasibility study stage to real world deployment) that have taken place in the last five years.

6.2 Ubiquitous Computing

"**Ubiquitous Computing**" is a new paradigm of information communication technology. This field, referred to as the "Internet of Things" in the EU and the "物聯網" in China, is receiving attention all around the world.

Ubiquitous computing is a technology with which computers and sensors that have been reduced in size due to recent advances in computer technology are embedded in various objects and places in our surroundings, and they communicate with each other and process information in a coordinated manner to offer useful information services for humans and to perform environmental control (Figure 6.1).

The term, ubiquitous computing was coined and popularized by Mark Weiser of PARC in the early 1990's [23], but the research in this area with the same notion had been conducted elsewhere before that. For example, an intelligent house called TRON House was built in 1989 in Tokyo [7, 20] under the leadership of one of the authors (Sakamura) in the TRON Project [13]. Some perspective on the early days of ubiquitous computing research at PARC, Olivetti, and in Japan is found in [2].

We are not going to give the full discussion of what ubiquitous computing is in this paper. We simply explain the essence that is necessary for the later exposition of applications.

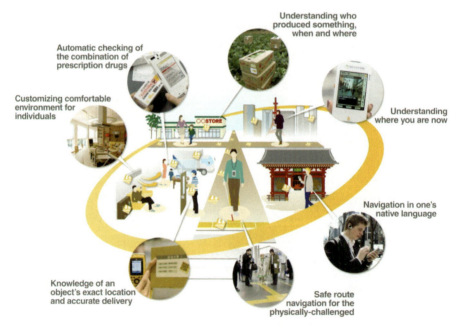

Fig. 6.1 The usage concept of Ubiquitous Computing Image.

In realizing ubiquitous computing as described above, the most important concept is **context awareness**. This means that a countless number of computers and sensors embedded in our surroundings recognize the real world situations so that they are used to offer advanced information services and to perform environment control.

Here, the easiest-to-understand "context" of the situation in the real world is what the object in front of us is and where our current location is, for example. If computers can automatically recognize such context, more convenient services can be offered to users than otherwise. (see Refs. [1, 5] for full context-awareness discussion.)

In order to realize such context awareness, it is necessary to **recognize objects and places reliably**. With the current technology, the surest and easiest method of recognizing objects and places is to assign a number (ID: Identifier) to the target which you want to automatically recognize, store the ID in a medium from which the ID can be easily and automatically recognized by a computer, and attach the medium to the object or place. For example, the most practical method is to print the ID as a bar code so it can be read automatically with a scanner, or to store the ID in an electronic tag typified by RFID (Radio Frequency IDentification) tags so it can automatically be read via radio wave. This is a very simple approach taken by uID architecture explained in the following. This is not the only approach, and there are obviously more complex methods for identification. But this article shows what uID architecture can do with this simplistic approach. The readers will find we can achieve a lot.

6.3 Ubiquitous ID Architecture

Ubiquitous ID architecture is a wide-area distributed architecture for retrieving information and services from objects and places in the real world that are identified by **ucodes** [8, 21].

Ubiquitous ID architecture has two assumptions. The first assumption is that various objects and places in the real world can be identified by numbers called **ucode**. To recognize this ucode automatically, bar codes, electronic tags, sensors, etc. (these are called ucode tags) where this ucode is stored are embedded in objects and places to which ucodes are assigned.

The second assumption is the establishment of an **always available network environment**, i.e., the **ubiquitous networks** of the 21st century, as the foundation. Of course, since there are places where a favorable digital communication environment cannot be established in the real world, the options, such as using local storage of pre-cached data, to operate in such an adverse environment has also been prepared.

Ubiquitous ID architecture consists of five components: (1) **ucode,** (2) **ucode tag**, (3) **Ubiquitous Communicator**, (4) **ucode resolution server** and (5) **ucode information server**. See Figure 6.2.

The method of acquiring information from ucode based on ubiquitous ID architecture is as follows. (1): **Ubiquitous Communicator** reads the **ucode** from a **ucode tag** using the automatic recognition technology. There are several ways of reading ucodes, such as automatic receipt of signals that active tags transmit, automatic reading of RFID tags, and bar code scanning. (2): **Ubiquitous Communicator** inquires the **ucode resolution server** as to the source

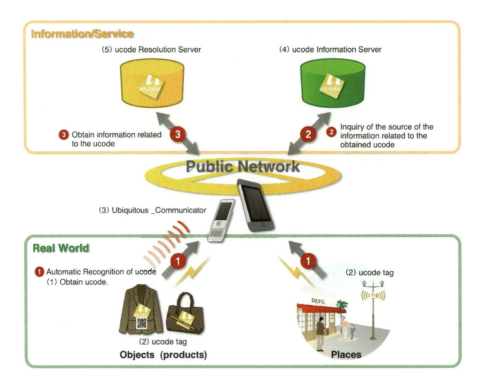

Fig. 6.2 Functional architecture outline of ubiquitous ID architecture.

(or location) providing information related to the read **ucode**. **Ubiquitous Communicator** can send context information to the **ucode resolution server** in addition to **ucode** itself. The **ucode resolution server** returns the source (or location) of the provided ucode information based on the ucode and context obtained from the **Ubiquitous Communicator**. (3) Finally, **Ubiquitous Communicator** connects to the information source or address of so called **ucode Information Server**, which has been acquired from the **ucode resolution server** and acquires contents and services.

The following components of ubiquitous ID architecture are important and we give additional explanation in the following sections.

ucode: Number which can be issued by anyone, any time, and for anything ucode is a number to identify objects and places in the real world. This ucode is an identifier system which identifies all targets of application. In addition to objects and places in the real world, ucode is used to identify abstract targets such as digital content, concepts, and meanings.

We must store ucode in some medium so that we can record ucode and read it back later, and attach the medium to objects and locations. The medium in which ucode is stored is ucode tag. While ucodes are uniquely assigned abstract numbers for identifying objects, and locations, ucode tags are physical media which are attached to objects and places in order to link the ucodes with the real world objects and places. The ucode tags can be print tags, or an RFID tag in which ucode is written, etc. Ubiquitous ID Center is **tag agnostic**. It does not use a single tag for all the applications. Many types of tags can be used as ucode tags. **Ubiquitous Communicator** is a terminal which works as a bridge between ucode and information by reading ucode and then subsequently retrieves services related to that ucode.

How does ubiquitous communicator find the service(s) related to a ucode? It uses, **ucode Resolution Server**, a wide-area distributed database server that manages the corresponding relationship between ucode and the location of servers (**ucode information servers**) that provide information, content and services related to the ucode. The ucode resolution server returns the address of the server providing information and services related to the ucode when an inquiry about ucode is made.

We explain some more details and features for some elements in the following.

6.4 Ucode

Ucode simply functions only as identification numbers in ubiquitous ID architecture. Attributes of objects and places to be identified are not described in the ucode number itself. But, ubiquitous ID Center does not prohibit such encoding of the attributes of identification targets into the ucode. (The application case of ucode by Geospatial Authority of Japan is a good example. See Section 6.19.1 later in this chapter.

ucode is stored in a physical medium, i.e., ucode tag. Inforamtion related to the object or the location identified by ucode is stored in remote ucode information server, and is fetched by using resolution server.

ucode can be issued by anyone any time for anything. ucodes can be issued for content and information which do not exist in the real world and for more abstract concepts as well as objects and places in the real world.

The ucode system is a 128 bit fixed length ($2^{128} = 340,282,366,920,938,463,463,374,607,431,768,211,456 = 3.4 \times 10^{38}$) identifier system. A mechanism to extend the code length in units of 128 bits has been prepared to meet the future demands so codes longer than 128 bits can be defined. The management and/or governance of ucode, such as maintaining uniqueness of code assignment, and lifecycle management of ucode (ucode is never reused) can be found in [9].

6.5 Features of Ucode

Compared to existing various code systems assigned to objects, ucode has the following advantages.

- ucode is a code to identify individual objects, not to indicate product types like a product code.

Product codes such as EAN, UCC, and JAN identify the type of product from each vendor. Therefore, the same product code is assigned to two packages of the same products. However, for ucode, different numbers are issued to individual packages even if they are the same product.

- ucode can be allocated to places, content, and concepts as well as objects.

ucode is the only code system that can identify objects, places, and content universally.

- ucode does not depend on application fields and business types.

ucode is not a code system to be used only in specific industries, for example, logistics. Therefore, ucode is very effective especially for services and item management across multiple industries and applications as well as for services that manage places and objects in the same system. ucode can act as glue to tie these applications together.

- ucodes do not contain meaning and are simple serial numbers.

This approach is effective especially for applications in which the meaning and nature of the objects and places to which ucodes are allocated change from moment to moment. Take a guardrail on a road, for example. Guardrails are products produced in a factory until they are delivered to a construction site. When they are installed at the side of a road, they turn into one component of the place. Lastly, from the time of their removal to disposal, they are treated as industrial waste. In this manner, even when the meaning (product/place/waste) changes from moment to moment according to the life cycle of an object, the ucode can simply continue to identify the item.

- ucode is tag agnostic
  ucodes can be stored in many type of tags (tags are described later).
- ucode can be used for secure application.

Ubiquitous ID architecture, the system for handling ucodes, has incorporated eTRON architecture which is the ubiquitous security framework [27]. Therefore, robust security and privacy information protection can be achieved.

6.6 Ucode Tags

ucode tags are the media for storing **ucodes**. The ubiquitous ID architecture is **tag agnostic.** Many types of tags can be used by applications. These include print tags such as bar codes and two dimensional bar codes, electronic tags without batteries such as passive RFID tags, and tags equipped with batteries which send IDs to terminals in push-style manner such as radio wave beacons (markers), infrared ray beacons (markers), and active RFID tags. There are differences of cost and physical characteristics of these tags.

- Cost of tags
- Readability of tags when they are placed on metal surface
- Readability of tags when they are placed on water-rich objects
- Readability of tags over a long/short distance
- Tags with a high level of security (tamper-proof, etc.)

A universal tag which satisfies all the requirements of applications does not exist. Therefore, rather than forcing the use of a single tag, Ubiquitous ID Center has established the tag certification system, based on which the most suitable tag can be selected according to the target to which tags are embedded or attached, and the environment of usage.

Ubiquitous ID Center certifies ucode tags in order to handle various types of ucode tags comprehensively as part of uID Architecture. A tag certified by this procedure is called a "Certified ucode Tag." As of December 2010, there are 37 kinds of certified tags including recently certificated Acoustic tag. For the details of certification system and the criteria for certification, please see [10] for an example of certification specification of a simple optical tag.

6.7 Ubiquitous Communicator (UC)

Ubiquitous Communicator is a terminal for obtaining ucode from ucode tags. It receives information services related to the ucodes and provides services to the user. The notable characteristic of UC is that it is a communication tool between ubiquitous computing environment and people. The ubiquitous communicator has three types of communications: "Communication with objects," "communication with people" and "communication with the environment."

There are many types of UC terminals: Many sizes and shapes which fit the needs of particular applications for which they are designed (Figure 6.3). See [19] for other types of UC terminals. There are also various kinds of ucode tag readers. In uID architecture, we don't restrict the type of tags to a single type of tag. So the development of a multi-protocol reader is important. New such reader has been developed (Figure 6.4). Interested readers are referred to [12]. There is now a newer commercial version of multi-protocol R/W.

Fig. 6.3 A few types of Ubiquitous Communicators.

Fig. 6.4 Example of Reader/Writer devices (left: Multiple protocol R/W with tags that are supported, right: R/W on a tip of a cane).

We feel that the smartphones such as android phones will, and should incorporate the reader/writer for RFID and other markers in the future. We are trying to influence some designs right now.

6.8 Ucode Resolution Server and Information Server

The ucode resolution server is a wide-area distributed database server that manages the corresponding relationship between ucode and the location of **ucode information servers** that provide information, content and services related to the ucode. When a UC makes an inquiry to the ucode resolution server about a ucode, the ucode resolution server returns the address of the **ucode information server** that provides information and services related to the ucode. Typically a web server is often used as information server, but any information provision server will do as the ucode information server as long as it offers information and services associated with a ucode.

Identifying all "objects" and "places" for many applications requires the ucode resolution server to manage an immense number of corresponding relationships between ucodes and content locations. The ucode resolution server consequently offers the mechanism in which multiple servers are distributed widely in order to manage such immense number of ucodes. This system manages many ucodes by a multi-layered tree configuration. The upper-level servers are assumed to be operated by internationally recognized organizations such as countries, international standardization organizations, trade associations, etc. and lower-level servers by enterprises or individuals. [6]

6.9 UID 2.0 — Realization of Richter Ubiquitous Computing World Based on ucR

Ubiquitous ID Architecture 2.0 is an extended version of ubiquitous ID architecture by incorporating meta information processing technology called ucR (ucode Relation) in the ucode resolution step. As a result, richer description of the real world and context-aware ucode resolution based on it can be realized. Interested reader are referred to [14, 17].

6.10 Ucode Usage Examples

Many initiatives that utilized "ucode" were carried out in FY2010 and in previous years. (In Japan, many organization's fiscal year (FY) starts in April and ends in March of the next year. Hence, FY2010 is from April, 2010 to March, 2011.) The infrastructure to connect objects and places by utilizing **ucode** has steadily spread. Some examples are enumerated below [18].

- ucode Application for Places
  General Location Information Systems
  Free Mobility Assistance Project (for helping the aged, physically-challenged, sightseers, etc.)
- ucode Application for Objects
  Food Traceability Systems
  Drug Traceability Systems
  Flashing Tag for correct shipment of drugs
  Pedigree and identification management of Thoroughbred Horses
  Electronic Medication Record

Common Pass for Construction Workers

Smart Houses (the world first smart house, and its descendents).

We pick only a few notable examples for exposition below. Note that the systems described below are the results of collaborations between government, private sector, and academia over several years. It takes time to develop a social technology infrastructure as envisioned by the ubiquitous computing paradigm.

We explain the applications that are possible by identifying locations first, and then applications for identification of objects.

6.11 Application for Places

The first example for location information-based system is Tokyo Ubiquitous Technology Project which has been going on since 2005 in the capital of Japan: This is a project in which uID architecture is used extensively. The applications developed in Tokyo Ubiquitous Technology Project share many characteristics of other regional developments which we touch briefly in the sections that follow.

The second example is *kokosil*, a location-based information system that serves as a generic system that has grown out of the early efforts as a platform for system integrators to build a location information service such as sightseeing guidance, restaurant guide, and other services.

6.12 Tokyo Ubiquitous Technology Project

The 'Tokyo Ubiquitous Technology Project' [25] is a trial in which the **Tokyo Metropolitan Government** has cooperated with **the Ministry of Land, Infrastructure, Transport and Tourism**, as well as local shopping districts with the aim of realizing an information provision service that improves the appeal and vitality of the district and makes it possible for anyone to walk around the district comfortably and safely. This was to be achieved through the use of ubiquitous computing technology, which allows necessary information to be accessed 'any time, anywhere, and by anyone.' This trial has been carried out in Japan's iconic shopping district, Ginza, and various other places in the Metropolitan area every year since 2006. It has been popular among the people who experienced the system, and the popularity has convinced the Tokyo Metropolitan government to continue its funding and its own use of the systems in providing services.

6.13 Tokyo Ubiquitous Technology Project in Ginza

A large number of content and services have been prepared in 2011 through cooperation with local merchants' associations so that more people including tourists from overseas can enjoy Ginza. The service area covers the entire Ginza area of 1 square kilometre. A terminal that can receive such services, including shop and restaurant guide, route guidance, guide to public facilities in the neighbourhood, and sight seeing guide in general, can be rent from various places in Ginza (including one in front of the Tokyo Metro Information Center, Ginza Street Guide, and hotels around Ginza). The content is available in four languages: Chinese (in both traditional and simplified characters), English, Japanese, and Korean. (See [3, 4] for the discussion of underlying guidance technology.)

Furthermore, services in FY2010 (April 2010 to March 2011) are also linked with *"kokosil"* (explained later in more details), a regional information portal site, which brings together a variety of information about the Ginza district.

6.14 Governance and Sustainability: Participation by Private Companies

Although the government offices were the initial promoter of the projects such as Tokyo Ubiquitous Technology Project in Ginza and similar trials across Japan, the feasibility experiment field has been opened to private companies with the aim of advancing the practicability and commercialization of ubiquitous ID technology, and each participating company is examining its own technology. This year 10 companies in 7 groups are carrying out experiments until March 2011 utilizing the ucode infrastructure installed in the Ginza area. Such participation by the private sector is deemed important and necessary for a long-term development and usage of location information applications based on uID architecture.

6.15 Metropolitan Government Ubiquitous Sightseeing Guide

There are other services in Tokyo aside from the one in Ginza. For instance, the guide of the Tokyo Metropolitan Government Building, which is in 'Michelin

Fig. 6.5 Metropolitan government Ubiquitous sightseeing guide (left, middle), and portable information system at Ueno zoo (right).

Green Guide Japan' and is visited by many overseas tourist. (Figure 6.5(left, middle)).

6.16 "Portable Information System," Ueno Zoological Gardens

At Ueno Zoo, Tokyo, the Portable Information System using Ubiquitous Communicators was started with the "Ueno E-Navigation Experiment" in 2005, and has been constantly expanded. (Figure 6.5(right)) The system offers information about animals such as the rarely seen behaviour of animals during night and mating calls that are heard only a certain time of the year, the daily updated message from the chief animal keeper, etc. aside from the general information. The terminals have become very popular among children. Furthermore, the trial operation of an electric cart, which is guided by ubiquitous radio markers, has been underway since June 2010 with the purpose of widening the range of visitors by making it possible for senior citizens for whom walking is difficult to enjoy watching the animals by moving easily (Figure 6.6(left)). The trial operation is scheduled to continue until the end of January 2011.

128 From Ubiquitous Computing to Ubiquitous Intelligence Applications

Fig. 6.6 Electric cart in the zoo (left, middle), and Hama-rikyu Gardens guide (right).

6.17 Hama-rikyu Gardens Ubiquitous Guidance Service

Hama-rikyu Gardens is a huge park consisting of cultural property gardens that have been nationally designated as special places of scenic beauty and special historic sites. A sightseeing guidance is carried out by using Ubiquitous Communicators instead of putting explanatory billboards in the gardens which tarnish the scenic beauty. (Figure 6.6(right)).

6.18 *kokosil*, Location Information Portal

Services providing town information including information on restaurants, sightseeing spots, and transportation facilities have spread rapidly in recent years. *"kokosil,"* a location information portal which has a particular focus on linkage with **uID architecture** has started operation in 2010.

kokosil provides services to a variety of terminals assuming two scenarios: At home and in town (Figure 6.7) At home, it is possible to carry out a preliminary study or submit a review on a PC. In town, it is possible to send street directions or store information to mobile terminals using **push** technology.

The location information service using **push** technology is the distinguishing characteristic of *kokosil*. The location information content registered in *kokosil* will be automatically delivered to the user while they are walking around the town carrying a mobile terminal with kokosil-support such as a Ubiquitous Communicator or a smartphone. By linking with the ubiquitous infrastructure, *kokosil* can provide finely-tuned services even indoors and in underground malls where GPS is not usable.

Fig. 6.7 Concept of kokosil.

kokosil provides various plans for visitors to walk around the town on such infrastructures. "*kokosil* **Tour Guide**," which is provided in the Tokyo Ubiquitous Technology Project, is also one of these.

6.19 Use of Ucode Location Information Systems Spread to Many Regions

Location information services and mobility assistance services utilizing "**ucode**" have spread to every region in Japan.

6.19.1 Intelligent Control Point and Location Information Code

We list this example in which the bit field of **ucode** is given a special meaning by an application developer: uID archirecture itself doesn't impose any meaning at all.

With the aim of realizing a "society where necessary location information is usable any time, anywhere, and by anyone," the Geospatial Information Authority (GIA) of Japan (http://www.gsi.go.jp/ENGLISH/)has installed "intelligent control points," surveyor's markers to which IC tags are attached (Figure 6.8).

GIA intends to advance the maintenance management and utilization of survey control points. The intelligent control point is positioned in "Basic Plan for Advancement of Utilizing Geospatial Information" adopted by the

Fig. 6.8 Intelligent Control Point with IC tag.

Specifications for the location information code

ucode : 128bit

version, etc.	Identification Code
	① ② ③ ④ ⑤

64bit

● Classification: 2 bits (4 patterns) *Latitude and longitude are displayed up to one decimal place every second
● Latitude: 23 bits ··················· North latitude: (+)0˚ to 90˚, South latitude: (-)0˚ to 90˚
● Longitude: 24 bits ··················· East longitude: (+)0˚ to 180˚, West longitude: (-)0˚ to 180˚
● Height (stratum): 8+1 bits ············· (The highest building: 160 floors) 256 strata and the middle of the stratum
● Number of identifications: 6 bits···· 2^6 = 64 pieces identifiable

Expresses the location to a precision of 0.1 second (approximately 3m)

Fig. 6.9 Structure of location information code based on ucode.

Cabinet in April 2008. To date, IC tags have been attached to approximately 20,000 triangulation points in urban areas nationwide and the maintenance of the intelligent control points has been carried out.

The location information code uses a **ucode** and the latitude, longitude, and height (stratum), etc. are encoded in the lower 64 bits of the code as shown Figure 6.9 stored in IC tag. This usage is an example in which a developer and users agree to a specific pre-defined meaning for the bit-fields.

GIS intends to use the intelligent markers to offer location-information services such as the retrieval of the original survey data, maps that show the ownership of the land lots nearby. GIS's own service will be complimented by commercial service providers who will use the markers as part of their ucode tag grids.

6.19.2 Using UID Architecture with Other Leading Edge Technology

Location information service using **ucode** can be combined with other leading technology to offer customized services to the users in many places.

6.19.2.1 Signage and augmented reality

An example is a use of digital signage system demonstrated at a shopping mall, LaLaport KASHIWANOHA (Chiba Prefecture) by YRP Ubiquitous Networking Laboratory and Mitsui Fudosan (Figure 6.10(a)left). This digital signage system aimed at providing finely-tuned information according to the particular user, location, time and circumstances of the user.

Display terminals called InfoScope that uses AR (Augmented Reality) to show the information about the surrounding area by overlaying such information on real-world images has been offered to public use in Yokosuka city by YRP UNL since June 2010. (Figure 6.10(b)) Coupled with tags that use visible light for communication [11], such AR terminal can show information about the tags in the scenary by overlaying it on the real scene and very useful, for example, in showing the location of attendants in event venues, sport stadiums, and such.

6.19.3 Regional Developments Outside Tokyo

Of course, other regions outside Tokyo Metropolitan area have adopted systems using uID architecture. Some photos from these systems are shown below.

(a)   (b)

Fig. 6.10 (a) Signage Application, (b) AR Application.

Fig. 6.11 Yukupon Card (left) and information provided after the card is scanned (right).

6.19.3.1 Miya sightseeing hospitality guide (utsunomiya city)

"Only now," "only here," and "only for you" information about Utsunomiya is sent by having a mobile phone read a **ucodeQR** (ucode embedded in QRcode) printed on a card called "Yukupon" that is distributed in stores (Figure 6.11). These coupons are often used to attract visitors to particular shops and restaurants by the local business community.

6.19.3.2 Yomitan village ubiquitous guide (yomitan village, okinawa prefecture)

Yomitan Village has established the "Yomitan Village Ubiquitous Guide" through the Yomitan Ubiquitous Village Construction Project, utilizing the subsidies for the Ubiquitous Town Concept Project and Promotion of The Use of Regional Telecommunications Technology granted by the Ministry of Internal Affairs and Communications. It has started since August 2010 (Figure 6.12). This system helps the visitors to the Yomitan village to learn the routes to interesting places and to use the unfamiliar local bus service easily.

6.19.3.3 Other guidance systems at historical sites

Tsuwano is an old town in the western part of Japan with its share of historical heritage. Sightseeing guidance system using uID architecture has been

6.19 Use of Ucode Location Information Systems Spread to Many Regions

Fig. 6.12 Terminal for Yomitan village ubiquitous guide (middle) outline of Yomitan village ubiquitous guide service (right) installed ucode wireless marker.

Fig. 6.13 Ubi-navi terminal (middle and left) using the service in Tsuwano.

offered there to guide the visitors to the famous sightseeing spots in town (Figure 6.13) [24].

Oyama road, an old highway walked on by people, goes through Kawasaki city. A local government office has tied up with a university laboratory to offer tour guide to historical sites along the road. ucodeQR printed on guidance panels can be read by many mobile phones in Japan to offer information services to the visitors to the area (Figure 6.14). It was found that visitors to the area now spend more time there by visiting otherwise unfamiliar historical sites along the road.

6.19.4 A Shared Solution: Ubiquitous Furusato Tourism

Sightseeing guidance and other location information services are well received by the local town planners for vitalizing regional economy. But the overhead of building the information provision system and maintenance individually are

Fig. 6.14 Panel with ucodeQR attached on a history guide of Oyama michi machi week.

the cause of headaches for municipal governments and regional commercial organizations.

As part of the "e-Regional Resource Utilization Project," the Japan Foundation For Regional Vitalization (commonly called Furusato Foundation) has since FY2008 operated the "Furusato Ubiquitous Common Platform" to unify management of the information (content) possessed by local entities such as local authorities and tourism associations for multiple local authorities to collaborate under common themes such as nature, literature, history, culture and festivals, and transmit the information (Figure 6.15). To date, 16 councils nationwide composed of approximately 130 municipalities have been sending out information using this common platform [26].

6.20 Application of Ucodes for Objects

There are numerous examples of assigning ucodes to housing, building materials, assets, etc. for management purposes. For applications for objects, we limit our applications to the following three which have used more than million ucodes to date.

- The management of historical information of houses.
- The Traceability of Housing components
- Cyber Concrete application.

(For other applications we can't cover here, see [18].)

Fig. 6.15 Website of "Furusato Ubiquitous," ucodeQR tag plate installed in a tourist spot.

6.21 Ucode Application in the Management of Historical Information of Houses

As the declining birthrate and aging population is becoming a bigger burden on society, and global environmental concerns and waste problems are worsening, the conversion from a consumer-driven society of "scrap and build" to a stock-oriented society of "creating high-quality products, properly looking after them, and using them carefully over a long period of time" is now an urgent issue in housing related fields.

In order to continue using a house for many generations, effective use of the past record of the house maintenance (historical information of a house) over a prolonged period is important in stages of the equipment upgrade, repair/remodeling, buying and selling etc. For that reason, in addition to building a system for passing on historical information of a house even after ownership changes, it is important to manage this information from the time of construction so that such information can be used later for remodeling.

As an effort to realize this objective, many organizations that specialize in managing historical information of houses have been established and services to assist house owners have been launched. In addition, the Historical Information of House Accumulation/Utilization Promotion Council has been established following a pilot program (2009). Their logo and name means a medical record for a house, figuratively speaking. (Figure 6.16) The Council is responsible for the distribution of identification numbers called "Common IDs" for the purpose of identifying individual houses uniquely as well as protecting personal

Fig. 6.16 Logo for "Iekarute" (House Medical Chart or Carte) which is a nickname for the historical information of houses.

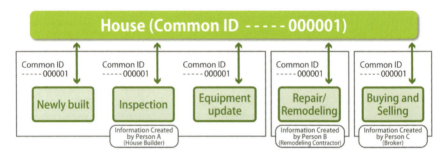

Fig. 6.17 Adoption of ucodes as IDs that uniquely identify houses.

information and privacy. The Council assumes the role of gathering information on houses separately managed by multiple information service organizations. Information service organizations and the Accumulation/Utilization Promotion Council manage historical information of houses by using "Common IDs" that uniquely identify houses as the key. The **ucode** has been adopted for this "Common ID." (Figure 6.17) "Common IDs" for 1.35 million houses were distributed. Use of the service has already begun.

For the details of the Historical Information of House Accumulation/Utilization Promotion Council, please visit their web page (http://www.iekarute.or.jp/).

6.22 Traceability Management System of Housing Components

"Who produced the housing components? What kinds of components were used? Who brought these components? And how were they used

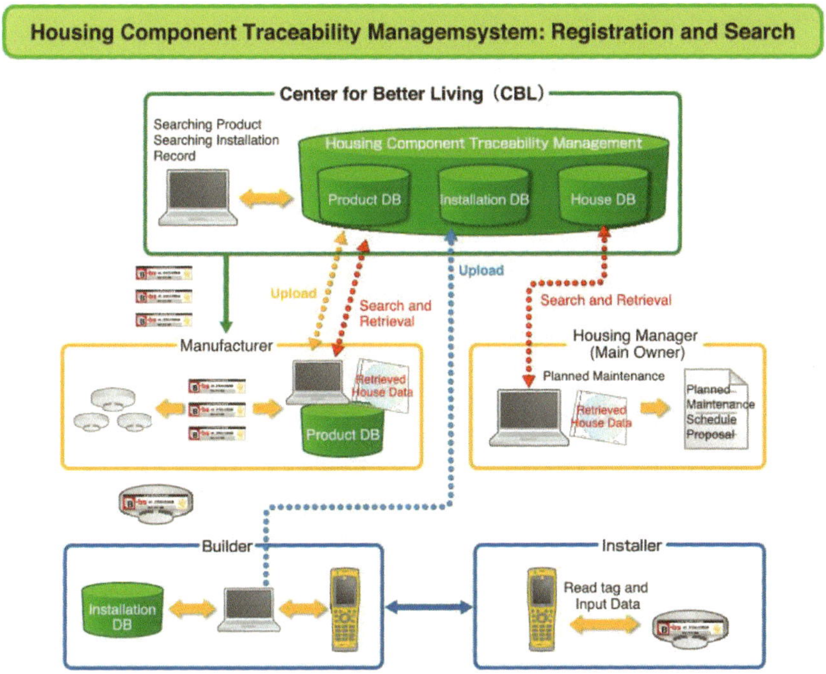

Fig. 6.18 Structure of traceability management system of housing components.

and inspected?" The Center for Better Living is operating "Traceability Management System of Housing Components" that can resolve those concerns of residents. (Figure 6.18). The operation of this Traceability Management System began in February 2006 with the focus on fire alarms, and information on the installation locations of approximately 1.6 million fire alarms for houses has been registered and managed with **ucodes** as of the end of November, 2010. In the case of fire alarm units, the units are tagged upon shipment from the factory. Installers use the ucode tag during installation to record when, where, and by whom the unit is installed. The future service includes the notice of battery replacement, and recall of defective units, etc. (See the action in [22]).

6.23 Cyber Concrete

Concrete, an important construction material, consists of a very large number of processes from material design to manufacturing, quality control, transportation/ delivery, casting (constructing), mould removal from the concrete, and curing. Its later maintenance management can be included in these processes. In addition, concrete has a unique feature where its properties and form change over time. Therefore, for concrete structures which are used over a long period of time, a great deal of effort is required for accurately and systematically organizing the properties of the concrete and circumstances during construction, its inspection history, etc. Moreover, it has been difficult to share the relevant information.

Sumitomo Osaka Cement regards the application of ubiquitous computing technology as very effective in the construction field for assisting the use of the structure "for a long period time" and "safely and without anxiety." Sumitomo Osaka Cement has introduced ubiquitous computing technology to realize well-defined responsibility/accountability, accurate information communication and sharing, securing traceability, and the construction of systematic utilization system and as a result, reliability related to houses/structures, etc. will be further improved.

6.23.1 RFID Tag Embedded in Concrete Test Specimen

Sumitomo Osaka Cement has introduced the use of RFID tags embedded in concrete slabs for quality assurance in its concrete test specimen management system. For fresh concrete, whether or not the target strength is met is inspected by applying a load to a test specimen (test piece) created from mixed concrete after the test specimen has reached the predetermined material age.

In the traditional concrete test specimen management system, capital intensive manual recording of test piece numbers and manual selection that was highly error-prone was a norm. Now, Sumitomo Osaka Cement has introduced the use of **ucode IC tags embedded in concrete pieces.** They can now be read and site information from the database is displayed. Since embedding these RFIDs in test specimens allows easy identification of individual test specimens, accurate traceability can be realized. In addition, by utilizing the system for entering test results and organizing data as well, test efficiency and accuracy can be improved.

6.23.2 Fresh Concrete Quality Assurance Information System

This application is a good example of an application made possible by combining different applications that span different application domains using **ucode** as glue.

Fig. 6.19 Tag for Cyber concrete (left: Sectional image, right: Scanned from outside).

The "Fresh Concrete Quality Assurance Information System" that includes transportation management was developed as an upgrade of the concrete test specimen management system (Figure 6.19). This system coordinates the fresh concrete quality control system and the fresh concrete shipment management system by linking information to **ucodes**. Furthermore, the system is extened to support transportation management when connected to an external network. Thus, all information related to quality assurance and shipment of concrete is organized in an integrated fashion. (See more details in [15, 16, 19]).

6.24 Summary

We have explained systems that are used in real world based on uID architecture.

uID architecture in its simplest form uses a very naive approach to context-awareness and identification. Very rich application can be built with such simplistic approach. We are moving to a more complex and richer approach based on uCR in the future.

Each of the application systems described faced the funding and development issues. Many have found a solution in the form of cooperation of government, private sector, and academia for the development, and a mixed support from the government and private sector for continuous usage and maintenance after initial development.

We are investigating a new application of SmartGrid that address issues related to eco-friendliness beyond a single house and in the context of communities, tows, cities and regions in FY2011. This again uses uID architecture. We hope the readers can learn something from these past and current systems to plan the future deployment of IoT systems.

References

[1] G. Abowd, et al., "Towards a better understanding of context and context-awareness,"*Lecture Notes in Computer Science*, 1999, Volume 1707/1999, Springer Veralg, pp. 304–307.

[2] B. Begole and R. Masuoka, "Search for eden: Historic perspective and current trends toward the ubiquitous computing vision of effortless living," *Information Processing Society of Japan Magazine*, 2008; 49(6), pp. 635–640. Original English version [Online] Available: http://www.parc.com/research/publications/details.php?id=6469

[3] M. Bessho, S. Kobayashi, N. Koshizuka and K. Sakamura, "uNavi: Implementation and deployment of a place-based pedestrian navigation system," in *Proc. First IEEE International Workshop on Software Engineering for Context Aware Systems and Applications* (SECASA 2008), Jul 2008.

[4] M. Bessho, S. Kobayashi, N. Koshizuka and K. Sakamura, "A space-identifying ubiquitous infrastructure and its application for tour-guiding service," in *Proc. ACM SAC 2008* (2008).

[5] A. Dey, "Understanding and using context," *Personal and Ubiquitous Computing*, 2001, vol. 5, no. 1, pp. 4–7.

[6] N. Koshizuka and K. Sakamura, "Ubiquitous ID: Standards for ubiquitous computing and the internet of things," *Pervasive Computing*, October–December 2010 (vol. 9 no. 4), pp. 98–101.

[7] D. Normile, "From Japan: Intelligence with Classic Style," *Popular Science*, 1990, pp. 58–61 (September).

[8] Ubiquitous ID Center, "Ubiquitous ID Architecture," 910-S002-0.00.24, [Online] Available: http://www.uidcenter.org/

[9] Ubiquitous ID Center, "ucode Management Implementing Procedures," Ver. 01.A0.10, [Online] Available: http://www.uidcenter.org/

[10] ubiquitous ID center, "Standard of ucode Tag Interface (Category 0)," 930-S211-0.00.02, [Online] Available: http://www.uidcenter.org/

[11] ubiquitous ID center, "Standard of ucode Tag Interface (Category 5)," 930-S216-01.A0.00, [Online] Available: http://www.uidcenter.org/

[12] S. Kobayashi et al., "A dynamic retargettable multi-protocol RFID reader/writer," *IEEE International Symposium on Ubiquitous Computing and Intelligence 2007*, vol. 2, pp. 340–346, May 2007.

[13] K. Sakamura, "TRON — Total Architecture," In *Proceedings of Architecture Workshop in Japan '84*. (August 1984), IPSJ, pp. 41–50.

[14] Y. Shigesada, S. Kobayashi, N. Koshizuka and K. Sakamura, "ucR based interoperable spatial information model for realizing Ubiquitous spatial infrastructure," *34th Annual*

IEEE Computer Software and Application Conference (COMPSAC2010), pp. 303–310, July 19–23, 2010. (Best Paper Award).
[15] WIPO patent application WO/2008/066177, "Concrete structure, method for manufacturing the concrete structure, and method for managing the structure and method."
[16] WIPO Patent Application WO/2008/066176, "Method for discriminating concrete specimen."
[17] ubiquitous ID center, "ucR format: ucode Relation Format," 940-S101-01.A0.00, [Online] Available: http://www.uidcenter.org/
[18] ubiquitous ID center, "Ubiquitous ID Technologies 2009," [Online] Available: http://www.uidcenter.org/
[19] ubiquitous ID center, "Ubiquitous ID Technologies 2011," [Online] Available: http://www.uidcenter.org/
[20] youtube Ubiquitous ID center channel, "TRON Smart House," [Online] Available: http://www.youtube.com/watch?v=7jPKEyM44GU
[21] youtube Ubiquitous ID center channel, "uID_architecture," [Online] Available: http://www.youtube.com/watch?v=mo8s2S78EOc
[22] youtube Ubiquitous ID center channel, "uID application," [Online] Available: http://www.youtube.com/watch?v=Y193LQUZqW4
[23] M. Weiser, "The Computer for the 21st Century," *Scientific American*, 1991.
[24] T. Ishikawa, "Ubiquitous Computing and Spatial Information: Toward a Ubiquitous Spatial Information Society", 2007 [Online] Available: http://www.jacic.or.jp/acit/3rd-pr_12.pdf
[25] Tokyo Ubiquitous Technology Project, [Online] Available: http://www.tokyo-ubinavi.jp/index_en.html
[26] Furusato Ubiquitous, "Furusato Ubiquitous" web page at http://www.furusato-info.jp/index.php?lang=en
[27] K. Sakamura and N. Koshizuka, "The eTRON wide-area distributed-system architecture for e-commerce", IEEE MICRO, Nov/Dec 2001, 21 Issue: 6, p. 7–12.

7

Technologies, Applications, and Governance in the Internet of Things

Prof. Lirong Zheng[1], Hui Zhang[2], Weili Han[1], Xiaolin Zhou[1],
Jing He[2], Zhi Zhang[3], Yun Gu[1], and Junyu Wang[1]

[1]*Fudan University, China*
[2]*China Electronics standardization institute (CESI), China*
[3]*KTH, Sweden*

7.1 Overview

The Internet of Things (IoT) is a vision of connectivity for anything, at anytime and anywhere, which may have a dramatic impact on our daily lives similar to the Internet done in past 10–20 years. It is recognized as an extension of today's Internet to the real world of physical objects, which is often associated with such terms as "ambient intelligent," "ubiquitous network," and "cyber-physical system." Its development depends on dynamic technical innovation in a number of important fields, ranging from fundamental microelectronic devices, sensor technologies to information and communication technologies (ICT).

The IoT has become a hot topic in China since Chinese Premier Jiabao Wen made a speech in Wuxi, where he called for the rapid development of the IoT technologies in 2009. Premier Jiabao Wen followed up with a speech on November 3, 2009 at the Great Hall of the People in Beijing, in which he encouraged breakthroughs in key technologies for sensor networks and the IoT. China has considered the IoT one of the key technology in its 12th national plan of 2011–2015. The equation: Internet + Internet of Things = Wisdom of the Earth, was widely referred and the IoT has become a hot topic for research and business investment in China.

This technological revolution of IoT brings many emerging applications and services, creating added value in market place. Several early-bird applications of the IoT in logistics, such as tracking, and healthcare have already been deployed in China. These technologies include architecture models, network technology, communication technology, discovery and search engine technologies, security and privacy technologies, and these applications include the smart grid, smart transportation, smart supply chain, and intelligent traffic.

7.2 Key Technologies

7.2.1 Architecture Models

There are two views for the architecture models: one is the vertical view which focuses on the technical implementation; the other is the horizontal view which focuses on the deployment and management.

In the vertical view, the IoT is characterized by comprehensive perception, reliable transmission, and intelligent processing. As is shown in Fig. 7.1, they correspond to the three-layer architecture: the sensing layer, the network layer and the application layer. Generally, the sensing layer realizes comprehensive

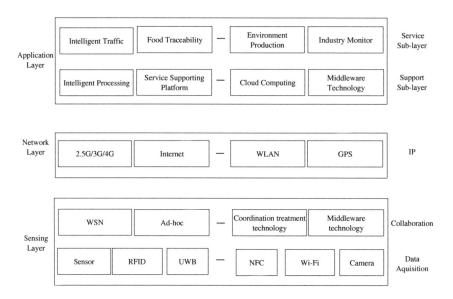

Fig. 7.1 The IoT architecture model in the vertical view.

perception by collecting real-time dynamic data through various sensors (including tags) while the network layer is mainly responsible for the reliable data transmission, relaying data acquired from the sensing layer to the application layer. Using distributed computing technologies, including cloud computing, the application layer performs massive data processing and intelligent analysis for the purpose of intelligent control.

We describe three layers in the architecture as follows:

Sensing layer: Two main functionalities, data acquisition and collaboration, are considered here. The event or state of "things" in the physical world such as temperature, concentration, and multi-media data has been perceived and acquired by sensing devices, such as sensors, RFID (Radio Frequency IDentification) tags, cameras and GPS terminals. The advanced techniques in this layer focus on the designs and implementations of new miniaturized sensors with low power consumption and high performance, the embedded technology, and short distance communication technologies, such as RFID, UWB (Ultra-wide Band), NFC (Near Field Communication). Collaboration technology applies for short distance data transmission, context awareness, massive information processing etc. It focuses on the hotspot technologies such as WSN (Wireless Sensor Network), Ad hoc network, in which the physical layer technologies (MIMO, OFDM, multi-hop etc.), data link layer protocols, routing protocols and relevant algorithms are the hot topics. Other technologies under observation include devices search and discovery, network edge expansion ability and the seamless integration with mobile communication networks. In the IoT system, security remains a hot potato. Perception nodes installed in unmanned environment can be easily destroyed. Sophisticated security technology is not available because of energy and cost limitations. As a result, how to achieve security and reliable transmission in the wireless sensor network or self-organization network is a dilemma.

Network Layer: In the IoT architecture, we are most familiar with this layer for we have used this system for many years and it has truly brought us great evolution and convenience. It is a heterogeneous network combing backbone, mobile communication network, WLAN, satellite communication network etc. The data acquired from sensing layer need to be transmitted safely and reliably through this layer. Although the network technology has come of age, there still exist problems to be solved to meet the new requirements of the

IoT applications: (i) Addressing: each "thing" in the IoT is mapped to only one address in the digital world. Due to the large scale of IPv4 in the existing Internet, conversion from IPv4 to IPv6 would be a long process. This situation brings up the compatibility problem. (ii) Network Integration: the IoT is a complicated, real-time heterogeneous network with the large scale. Multiple heterogeneous terminals are intensively deployed in place. The strategies for network integration are fusion and collaboration. Fusion refers to integration of various heterogeneous networks including the mobile network, Internet, PSTN, Wi-Fi, Bluetooth and GPS. In short, it forms a network containing all. Collaboration means the integration of "personality" in different systems, specifically referring to access subnets to realize coexist, competition, and cooperation to meet business requirements. (iii) Resource Management: massive, dynamic, dispersible data have to be stored, transmitted or processed. These requirements lead to both the access networks and the core network to design a new topological structure or interactive way to improve the network resource utilization efficiency and throughput. At the same time, the networks have to be robust and intelligent enough to adapt the dynamical situations.

Application Layer: This layer includes the Support sub-layer and the Service sub-layer. After storing, processing and analyzing data intelligently, the Support sub-layer delivers the results based on users' requests. As a vast array of data involved, the system integrates distributed computing technologies, such as P2P (Peer to Peer) and cloud computing, which both facilitate intelligent analysis and processing, decision making, and enhance the capacity of information processing in the IoT. With open application interface upwards, the sub-layer masks differences among various accesses in the lower layers shown in Fig. 7.1. General device finding, addressing, routing, QoS (Quality of Service) service control, billing, management and security control are supported as well. Those functionalities are implemented by middleware, object name resolution, etc. Research on this sub-layer focuses on massive data storage, intelligent information processing, data mining, structure supporting, cloud computing and service oriented, network enabling, and cooperation application in the IoT. Based on the Support sub-layer, the sub-layer of application services provides interfaces and platforms from an extensible service structure. This layer mainly orients information services, while supporting individual public services. Typical applications include monitoring in security, and disaster, as well as intelligent household appliances and vehicle scheduling. Openness and standardization are imperative for scale effect, as a result of no standardized management platform.

The three-layer model introduced above has certain significance to understand the framework or key techniques of the IoT. However, it cannot express the whole features and connotations. It needs to be added and improved with the development of the IoT technologies. Some researchers have already put forward a five-level architecture, including the business layer, application layer, processing layer, transport layer and perception layer. No matter what the changes are, the architectures remain essentially three keywords: perception, transmission and processing.

In the horizontal view, the IoT consists of different domains. The criterions to identify a domain include different countries, different states, different industries, and different technologies. This view can reduce the task burden of deployment and management. Each administrator and manager can reduce their range to their familiar systems and areas. But this view will lead to other problems: how to manage the different domains, and how to integrate the different domains. Because each domain is deployed different technologies, policies, laws, and even locations, a big problem will be arised when integrating domains.

7.2.2 Network and Communication

The IoT contains various kinds of local networks. WSN, which is attractive due to its convenience of deployment and the potential of autonomy, provides a broad research space. Low cost routing discovery would be the key technology, because WSN is highly limited on computing and power resources. Once the battery of a node is exhausted, it cannot be activated, and it is usually impractical to replace the battery. Though research has been conducted on passive communication, the battery is still necessary in most QoS related applications. Thus, WSN sensor power allocation and WSN autonomy routing discovery are under consideration.

Cluster provides a possible approach for routing discovery in a considering scalable WSN network. It can save routing expenses and extend the network life time. The strategy of cluster algorithm is important in the cluster to maintain low cost and QoS. Furthermore, a different application requires a specific cluster algorithm. Density and mobility may vary greatly in different applications of the IoT. These factors are important in determining which algorithm to use.

Network coding provides another approach for the scalable WSN network routing. It can improve the system throughput, balance the network load and achieve robust routing. The network coding is originally raised to solve the broadcast problem, while applying it to WSN is also an important issue. WSN significantly differs from the traditional communication link in that the WSN may experience dynamic network topology and routing in its life time. The tremendous WSN nodes require a scalable network coding strategy. The stability of the network coding is also concerned. WSN may be working in poor SNR and bit error rate, while in the network coding high BER condition is required.

WSN is a resource limited network, but the network coding increases the complexity of intermediate nodes. The ad hoc network, where all network nodes can be intermediate nodes, especially hinders the appliance of the network coding. The network coding may also introduce additional transmission delay, which may be critical in some real-time applications. More research on the practical network coding is required.

When deploying the IoT, it needs to be remembered that the UHF frequency band and the MSI frequency band, used to major the IoT technologies like RFID or 802.15.4, are both too crowded. Mobile communication and WLAN are sharing the same spectrum, which may lead to severe interference. Exclusive frequency allocation is difficult and often not enough for application. However, researchers revealed that many authorized frequency outside UHF and MSI is not fully utilized. Thus the progress in cognitive radio research provides the idea of realizing spectrum reuse through intelligent and dynamic management. There are still difficulties in applying cognitive radio to ad-hoc networks, as multiple detection and transmission are employed. Effective and stable schemes for ad-hoc networks of the IoT are essential.

The achievements in cognitive radio enable many interesting applications. The achievements of cognitive radio in power and spectrum management can also contribute to the IoT. As the IoT is often a self-organized system, in which system nodes are required to manage themselves without the knowledge of the whole system, the transmission frequency band and signal power has to be decided according to local knowledge. Research achievements in cognitive radio can then be applied to avoid collision and waste of resources.

In scalable WSN, all nodes share a spectrum band and it is up to the node to decide the transmitting parameters in ad hoc situations. In this context, cognitive radio may become even more important. Nodes are required to coordinate with each other and also with the environment. Strategy like CSMA/CA is usually employed to cope with possible collision and hop to another frequency band. In the IoT context, the resources wasted in collision cannot be ignored. A non-collision solution is required to fulfill the frequency reuse. Cognitive radio may enable nodes to detect the radio environment, finding a solution towards frequency space time reuse avoiding collision.

Because the IoT involves various network technologies and system architectures, achieving heterogeneous networks convergence among all networks is an important issue. Current wireless network technologies, including RFID, 802.15.4 and Wi-Fi, may find a communication solution in an IP based network, which is also commonly recognized as the most practical solution toward the full vision of the IoT.

RFID is the most widely used short range radio technologies. In China, RFID has been successfully used for Chinese ID Card, liquor anti-counterfeit, library management, appliance management, and Shanghai EXPO ticketing, etc. A typical RFID system consists of a tag, a reader and a database. Because of the low cost, passive RFID is more popular than the active RFID and semi-passive RFID. ISO 18000-6C is a widely-accepted standard for ultra high frequency (UHF) band RFID applications, and ISO 14443 is a widely-accepted standard for high frequency (HF) band RFID applications. The key technologies of RFID are power management, security and privacy protection, low cost integration of RFID and sensor, effective and robust searching technology, and international standardization.

As the radio frequency identification (RFID) systems grow in size to thousands of tags for many applications, transmission collisions, such as tag-to-tag collisions, reader-to-tag collisions, and reader-to-reader collisions, may occur when there are many readers and plenty of tags within close vicinity. A tag-to-tag collision occurs when multiple tags respond to a reader simultaneously. A reader-to-tag collision occurs when a tag is within the interrogation zones of multiple readers and more than one reader attempts to communicate with that tag simultaneously. A reader-to-reader collision occurs when a reader, which is receiving a tag response, is interfered by stronger signals from one or more neighboring readers operating at the same frequency simultaneously. In our opinion, tag-to-tag and reader-to-tag anti-collisions are more important, since the MAC schemes that can solve reader-to-tag collisions can solve reader-to-reader collisions as well but not vice versa. Broadly, MAC schemes can be categorized into space division multiple access (SDMA), frequency division multiple access (FDMA), code division multiple access (CDMA), carrier sense multiple access (CSMA), and time division multiple access (TDMA). TDMA schemes and TDMA-based combined schemes constitute the largest group of anti-collision protocols. In future RFID systems, combined schemes in MAC layer can further improve the performance.

IPv6 provides a scheme of interconnecting everything and every network nodes. An IP based scheme provides a global unified address allocating and routing solution which achieves the communication among different systems. It uses the packet switch mechanism, which is important for the IoT in meeting different business requirements and the well-developed network technology on the IoT network, as well as the current Internet infrastructure to realize global communication.

Because the IPv4 address space is using up, IPv6 is essential. However, applying IPv6 to the IoT still has a long way to go. Further research topics such as IPv6 head compression, mobility support, security and QoS are required on when applying IPv6 to the IoT business. For example, IPv6 head contain 32 bytes of addresses information, which is obviously too long for an IoT packet which is usually 20 to 50 bytes. Additional research on compression IPv6 head information without losing IP routing ability is needed.

The IoT applications are often highly emphasized on real-time requirements. For example, in the smart grid, if a data packet cannot arrive on time, a failure could occur. This may leads to the failure of corresponding physical system and tremendous losses. The IPv6 scheme should take the real-time requirement and other possible strict QoS requirements into consideration. The QoS assurance is important because the IPv6 scheme implies to use of the current public Internet infrastructure for wide area communication. The performance of a public network is sometimes difficult to be predicted and may lay serious impacts on QoS. Practical solutions are required for an IP based scheme.

It is also worth mentioning that heterogeneous networks convergence requires additional attention on QoS control. A communication link in the IoT applications may consist of a local WSN network, a wide area access like 3G/4G communications and an IP network routing switch. In China, TD-SCDMA system is already put into practice, and the 3GPP TD-LTE communication technology is under construction. Those mobile communication systems would afford abundant access resource for the IoT applications.

Vertical handover in heterogeneous networks context may provide the improvement in system performance. The IoT nodes may switch to a different access to avoid an access failure. This can be used to adapt public networks to the IoT applications, since the current public networks are not designed for the IoT real-time applications.

The communication technologies in the IoT network are developing very quickly to meet the requirements of the connections among physical world "things" and "humans" these years. Issues such as adaptability to the changing environment, access architectures, efficient power communication systems, etc. are being studied by researchers all over the world. The high density of the mobile devices and the communication between "things" need new methods and algorithms to solve the interference problems, combat bad wireless channel conditions, and improve the system throughput.

The object of communications in the IoT is mainly to achieve the interaction among the physical world, and digital world. Besides, new related communication technologies are as follows:

Interference Mitigation:

Because the IoT requires a high density of physical world devices, interferences are inevitable, such as multiple access interference and intersymbol interference. An efficient spectrum spreading technology — IDMA — is put up for solving this problem. The problem has been worked out through multiplexing by using different interleavers. Moreover, the technology increases the system capacity by allowing more information exchange and even acts as a solution to frequency spectrum allocation in the communication system.

Multihop Relay:

Multi-hop relay has been regarded as a new technology in the IoT relating to wireless backhaul networks, user cooperation networks and sensor networks. Multi-hop relay includes orthogonal multi-hop systems and non-orthogonal multi-hop systems. The latter one will perform a higher theoretical system capacity. The application of multi-hop relay will increase system throughput, enhance network coverage, and combat bad wireless channel conditions.

Various communication requirements and hardware upgrades problem:
The hardware platform has the characteristic of compatibility, scalability and interoperability. A new technology named software defined radio (SDR) has been used in the IoT. By using this technology, different communication functions can be achieved and the complexity of hardware upgrades can be lowered.

7.2.3 Discovery and Search Engines

7.2.3.1 Describe a thing

The Thing Description Language (TDL for short) is proposed to describe the basic units, which we notate as Things, in the IoT. We propose that a thing in the TDL consists of properties, relationships, behaviours, policies, and environments:

- The properties include the identity of a thing. But the identity could be optional. That is, a thing in the IoT could have no identity, because this phenomenon of no identity is common in the physical world. Beside the optional identification, the properties include other basic information of the thing, such as a lot number, and expiration time.
- The relationships are the links among things. Similar to the social network in the Internet, we argue the IoT also has social network where things has their friends, ancestor, and offspring. The things can also connect others through manufacturing, production, sale, living, and business. The relationships will express the above information.
- The behaviours are the interfaces and their definitions.
- The policies describe the interactive strategies when a thing cooperates with other things and the environments.
- The environments describe the features of in which a thing can live.

The language can describe all things in different environments in the IoT. These environments include tags, sensors, or back-end servers.

Currently, EPCglobal proposed the Physical Markup Language (PML for short) to describe the properties, processes and environments relevant to a RFID tag. In the architecture proposed by EPCglobal, the information described in PML is stored in an EPCIS server; and an ONS server will map the relationship between a RFID tag and the PML information. But we argue the information specified in PML is not sufficient. For example, the PML does not consider behaviours and policies of a thing.

Different from the PML, TDL will include more information. Particularly, TDL can describe an active tag or sensor in the IoT.

7.2.3.2 Discovery and search engine in the IoT

The IoT consists of many distributed and decentralized resources which are provided and required by different users and organizations around the physical world. Discovery and search services should be applied as soon as possible to meet the growing needs of gathering complete and accurate information and things in the IoT.

The standards of SOA provide specifications of UDDI (Universal Description Discovery and Integration) to help distributed web services cooperate. In UDDI, a service provider can register a service, and the service can be discovered by a service requester; then the service can be integrated into the business logic of the requester. But we think this technology is not enough for the IoT due to the decentralized governance, mobility of things, energy limitation, and performance limitation, etc.

As a thing roams through the physical world, the IoT has some specific features and limitations. In the first place, like most networks, it contains a wealth of information, which is offered by distinct users. Different from current networks, the description language in the IoT is well organized and semantic, which may contribute to supporting more efficient searching than the Internet. Last but not least, because the information in the IoT is important and even confidential, security and limitations must be taken into account. The discovery and search services must comply with the rules or laws set by the resource providers (including industry organizations and countries).

Considering these features, our goals is to choose and manage data and search algorithms to help provide a secure access to gathering complete and accurate sets of information among the large amounts of distributed resources from different organizations.

Search algorithm

In order to achieve the goal, we need to find suitable search algorithms. Advanced search algorithms like best-first search, stochastic search and simulated annealing, which are popular in artificial intelligence, but ignore the features of the IoT (the IoT is well organized instead of data accumulation).

What we propose to use is the P2P system. P2P (Peer-to-peer) is a concept widely used in distributed computing and web services, which considers the characteristic of the network. It can be suitable to be applied to the discovery and search services mentioned here.

First, for the purpose of implementing P2P to the IoT, we separate the whole IoT into several domains. The separation can be decided by the countries (e.g. USA, UK, and China), industries (e.g. food, entertainment, and livestock), organizations or the combination of any of them. Ideally, a thing in the IoT should be located at only one domain. However, in the physical world, a thing may be in several domains or cross from one domain to another one. For example, a bottle of milk may be both in the dairy industry and the retail products industry. For this reason, some domains of the IoT may intersect with others. The system can put up with this but should avoid it to the greatest extent.

Structures in each domain

After the domain separation, a data structure in each domain should be considered since it is closely related to the complexity and efficiency in the searching process. So as to achieve as good of result as possible, a hierarchical structure is purposed here. The concept "ontology" is exactly suitable to provide such a structure. Ontology is the structural framework for organizing information and is used in artificial intelligence, Semantic Web, systems engineering, software engineering, library science, and information architecture as a form of knowledge representation about the world or some parts of it. It is a formal representation of knowledge as a set of concepts within a domain and the relationships between those concepts. It is used to reason about the entities within that domain, and may be used to describe the domain. Using this concept to manage the things in each domain will help us discover the complete and accurate information in a limited time because of the well classification. For instance, in the pet domain, if the user looks for a cat born in 2011, then the system can trace the qualified cat through animal to mammal, mammal to feline, and from feline to cat, and then search the cat by the birth year.

This kind of data structure can save a lot of time when the amount of information is huge. However, in some cases the classification might cause confusion in how we should divide the things into hierarchical classes. In these conditions, simple data accumulation is allowed. However, just like the domain separation, we should avoid them and try our best to establish a good data structure in each domain.

According to the horizontal view, many domains with specific laws and policies will run in the IoT. When a user launches a request for something, the system asks all the domains (defined before) for a suitable answer. Each domain accepts the request and searches a qualified item or description in its local IoT with the limitation of its own rules or laws. After that, each domain sends a result to the system and the system gives the user its answer by intersection or distinction of these results. For example, a person wants to find a book and send a request to look for a book with his requirements. The request may be sent to each domain in the IoT. Some domains might ignore the request, thinking the user lacks rights under their laws. Others may accept the request and search for the suitable book. The system may merge all results from the domains and give an answer to the user.

The application of P2P algorithms and hierarchical structures in the discovery and search services take the laws and rules of each countries (or organizations) into account as well as manage the tremendous resources, which solves the basic problem of the IoT and is promising in the development of the IoT.

Search Engine

In order to support the discovery and search services described in the former part, the search engine in the IoT should be strong and flexible enough to face the condition that billions of search requests need to be handled in time, which is similar to some web search engines meet present. So the work mode of present web search engines may be useful in promoting the development of search engines in the IoT.

There are hundreds of web search engines nowadays. Some of them just provide search in a limited domain while others search through the whole Internet. Although they are different in the search range and ability, their search methods are similar. Almost all use spider to crawl pages in the Internet at first. They then do some pretreatment to these pages, such as extraction of keywords, removing duplicate pages, segmentation etc. When a user's request approaches, the search engine find pages which matches the keywords in the database.

The work mode of the IoT's search engine is similar to that of web search engines in a certain extent. Each domain manager (may be a server or human admin) scans all the things in its domain and sends the result to the search engine cache. The things can be registered when the tag was issued as well as passively scanned. After the first step (index things into cache), what the engine need to do is wait for a request and extract keywords of it. After obtaining, the engine search relates information or things in the cache. If there is suitable result in the cache, then the engine combines the results and returns it to the user. Otherwise, the search engine asked all domain managers for the searching thing (or information). And all the domain managers scan all the things in their domains and send a result to the search engine. The engine combines the results and response to the user. However, after such search (not find within the cache), the search engine refresh its cache with the latest scan results from the domain managers.

Although the work mode of web search engine and looks similar, there are still some distinctions between them. First, in the keywords extraction, the IoT search engine has to put in more effort since it must provide enough information to achieve an accurate and complete search. Second, after a search in the cache, if the engine cannot find a satisfied result, the web search engine will give up while the IoT search engine asks domain managers for a new scan. Finally, we have higher expectations for the IoT search engines than the web ones because the former has unified data structure and will be effectively organized. Senior searches, such as history tracking, dynamic information, are expected to be achieved on the IoT search engine.

7.2.4 Security and Privacy

Security and privacy are two of the important issues in the IoT, especially, when the IoT is widely used in our physical world, and many living processes, such online payment, transportation, will depend on applications of the IoT.

Here, we will discuss two issues from protected assets, threats to assurance techniques. In the assurance techniques, we propose some practical concepts, including multi-profile assurance, evolving security for the IoT.

Protected assets

In a typical the IoT application, the features and key assets include sensors (including tags), the communication channels, access points (including readers), and back-end systems based on the Internet. Although the back-end systems are the important part of a typical the IoT application, the security and privacy issues and their solutions are similar to the traditional ones in the current Internet. The possible significant difference between the issues in the IoT and the Internet is that the privacy issue in the IoT could be more important than the one in the Internet because the data stored in the IoT is nearer to the private information of a person, e.g., the living blood pressure. As a result, the privacy issue is prompted while the IoT is developing and continuously researched.

In addition, with the development of sensors (including tags), active sensors (especially tags) will be manufactured and used, e.g., a medical tag can be planted into a body and drugs can be actively injected against diabetes if conditions are met. Thus the environment (e.g. the body) containing the tag is also the protected assets.

Threats

Since the IoT applications are widely deployed, the threats to the IoT applications are also pervasive. The threats, we argue, consist of four kinds: Sensors (including tags) oriented, communication oriented, access points (including readers) oriented and environment oriented.

The sensors (including tags) are the most important assets in the IoT applications, especially when the mobility and portability are the basic physical features of the sensors (including tags). *First*, people including adversaries can touch the sensors. Thus the physical attacks, including theft, loss, destroy, must be considered when we set up the IoT applications. *Second*, the confidentiality and integrity of data must be protected. That is, the data stored or gathered in sensors (including tags), e.g., deposit balance or living blood pressure, are very sensitive, thus the reading and modification to these data must be authorized even audited. *Third*, the integrity of codes in the sensors (including tags) must be protected. That is, once the codes stored in the sensors (including tags) are modified or bypassed, any protection based on these codes will be un-trusted, thus compromising the nodes. *Fourth*, the threat to the availability of sensors (including tags) could happen when a sensor (or a tag) cannot work though it is not removed legally. This threat could happen when some synchronization information stored in the sensor (or the tag) is modified. *Fifth*, the Sybil attack where a fake sensor declares it is legitimate will seriously threaten wireless sensor networks because the fake node could hijack communication channels, forgery messages, etc. *Last but not least*, fake sensors, (including tags), covert channel, and side-channel attacks should be big threats when large-scale sensors (including tags) are deployed.

The communication between sensors (including tags) or between a sensor (or a tag) and an access point (or a reader) could be eavesdropped, interrupted, delayed, or modified.

The access points (including readers) usually connect the Internet and the sensors (including tags). *First*, the phishing attack could be the big problem. That is, an attacker could deploy a fake access points (or a fake reader), and lures a sensor (or a tag) to transmit its sensitive message to the fake one. *Second*, the integrity of codes in the access points (or the readers) could be attacked. Thus, the access points (or the readers) could be compromised, and then the data, e.g. secret keys, stored in the SAM (Security Access Module) could be leaked.

Environments where the sensors (especially active tags) work must be protected because the activities of the sensors (especially the active tags) could destroy the environments. In the threats to the environments, the adversaries are the sensors (especially the active tags). For example, if the active tag is planted in a body and drugs are actively injected, if a mistake occurs, either unconsciously or maliciously, the body would be harmed. The problem could be led by the breaking of the Principle of Least Privileges, where the sensors (especially the active tags) are authorized the permissions as necessary as possible.

In a word, the threats to the IoT seriously block the development of the IoT. Also the novel threats could appear when the IoT is developing. Thus, the relevant assurance techniques are developed to defend them.

Assurance techniques

Similar to other Internet-based systems, assurance techniques for the IoT applications include authentication, access control, and audit techniques. In addition, cryptography is also a key technique. When these techniques are used and deployed in the IoT application, the biggest challenge in protecting the assets is to use existing or novel techniques and to protect the security and privacy of the information in the IoT with the tough restrictions of *performance and cost*.

To protect the sensors (including tags), researchers proposed many light, even ultra-light weight cipher algorithms and authentication protocols. The performance and cost are two critical challenges because the massive volume sensors (including tags) will be widely deployed. Any redundant design will reduce the performance and increase the cost. As a result, traditional cipher algorithms, such as AES, RSA, cannot be deployed in the large scale. Researchers are trying to improve the performance and reduce the cost of the algorithms and protocols, including though proposing novel ones. With these cipher algorithms and authentication protocols, the identities of the connected sensors (or tags) will be authenticated. That offers the basic assurance for the confidentiality and integrity of the data and codes stored in the sensors (or tags). In addition, the Sybil attack can be defended and the fake nodes can be recognized.

Second, researchers proposed access control protocols to more finely protect the sensitive data stored in sensors (including tags). Generally, the mechanisms in these access control protocols are very simple but efficient. E.g. the sensitive data is protected based on a password/passcode-based protocol where once a request includes a legitimate password/passcode, then the sensitive data can be accessed.

Third, audit for any access to the sensors (including tags) is a necessary safeguard. With an audit mechanism, the analyzers can find the potential flaws in the implement algorithms, protocols, and products. However, the volume of audit data could be the big problem. The problem is two folders: On one hand, the storage in a sensor (including tag) is usually critical; thus, it is hard to store the access logs for audit in the sensor (including tag). On the other hand, it is also hard to set up a central server to store all access logs due to the huge-volume tags, integrity and privacy issues of these logs; thus, we argue the audit should be researched as soon as possible.

Fourth, physical safeguards, such as tamper resistance and the *kill* command, are researched and deployed. Tamper resistance can protect the confidentiality of codes by a self-destroy program. And the *kill* command, which can disable a tag, is a standard command in an EPCglobal-compliant tag, therefore protecting the user from the pervasive tracking.

To protect the communication in the IoT, efficient cryptography algorithms and protocols are used. Usually, the messages or data will be encrypted by a pre-shared key or a negotiated random key, then sent though a public channel, and decrypted by the relevant key. This way can efficiently defend eavesdropping and modifications. In addition, developers usually insert some synchronization codes or fresh numbers to defend the delay attack.

To protect access points (including readers), the light even ultra-light cipher algorithms and authentication protocols are deployed. Due to the connection with the Internet, the access points (including readers) are also required to be deployed with standard cipher algorithms and authentication protocols. Using the authentication protocols, the legitimacy of an access point (or a reader) can be identified, and thus defending sensors (including tags) from the phishing attack. In addition, to mitigate the threat from the integrity of codes embedded in the access points (including readers), we can design an architecture, where the secret data will not be stored or transmitted in the access points (including readers) in plaintext.

To protect the sensitive environments, e.g. bodies, standardized behaviour specifications for active sensors (or tags) must be designed and enforced.

In a word, to mitigate the threats from the each part of the IoT, the assurance techniques are developed and deployed for each part. However, the restrictions of the performance and cost for the IoT is tough, thus the current techniques could not mitigate some threats under the acceptable range of users. We therefore introduce the evolution of the assurance techniques.

Evolving Assurance Techniques

Due to the rapid development of micro-electronics technology, the performance of a unit cost rapid increases. As a result, some algorithms and protocols which cannot be implemented in the current sensors (including tags) would be implemented in the near future. We use the cipher implement in RFID to introduce this concept as follows.

According to our investigation, as the semi-conductor industry has been developed according to Moor's Law, there is more chip area available on tag chip for security enhancement. For instance, the capacity of a 0.1 mm^2 tag chip is about 1,800 Gate Equivalent (GE) for digital circuits in 0.35 micron technology, but the capacity changes to about 8,300 GE and 19,200 GE when the technology transfers from 0.35 micron to 0.18 micron and 0.13 micron technology respectively, though there are some differences among different foundries and different library providers.

In the RFID field, aimed at one or several security threats, many security mechanisms and authentication protocols have been proposed in the literature. The idea "block tag" was presented by Juels et al. to prevent the unauthorized tracing. Weis et al. proposed the cryptographic privacy enhancing technology based on hash-lock for the first time. Juels et al. introduced an HB+ protocol, which made use of the hardness assumption of statistical "Learning Parity with Noise" (LPN) problem. During the past few years, low-cost implementations of standard symmetric-key cryptography algorithms have been reported to build strong security protocol, such as Tiny Encryption Algorithm (TEA), International Data Encryption Algorithm (IDEA), and Advanced Encryption Standard (AES). There are a few reports on implementations of asymmetric-key cryptography algorithm for RFID, such as *Elliptic curve cryptography* (ECC). Hummingbird is a new light-weight algorithm proposed by Revere Security research team targeted for low-cost RFID tags. Hummingbird has a 256-bit key size and 16 bytes block size. Different from the normal symmetric crypto algorithms, the encryption process of hummingbird consists of an internal state initialization and a block encryption. Since the initialization process only needs to be done once during one communication, Hummingbird has an advantage over the normal symmetric algorithms when the plain text to be encrypted is very long. In the foreseeable future, the RFID tag chip cost is expected to be lowered by more than 20% per year because of the technology development, and stronger cryptography algorithm, even the asymmetric algorithm, will be appear in RFID tags. On the other hand, power consumption will become the dominant factor for the introduction of assurance technologies for RFID tags. Low power technologies and power efficient technologies will be the decisive technologies for secure RFID tag chip design in the future.

Security Management and Education

Security management first includes how to configure appropriate techniques to protect the appointed applications. As we all known, not all threats will happen in an IoT application, and not all threats in the IoT application must be mitigated; thus, we design different profiles for the applications. The profiles are hierarchical. That is, a high level profile could include all assurance techniques of a low level profile for a class of applications. This concept is similar to the one in Common Criterion.

Second, policy-driven management for security and privacy would be deployed in a large scale, due to the huge volume and complexity of the IoT. Thus we propose a standardized policy language and enforcement framework, which is similar to the one in the IETF/Distributed Management Task Force (DMTF) policy framework, will be motivated to be developed.

Third, the laws and boundary will be problems. Because the IoT applications could be deployed in the physical world, and the policies and laws in different nations, states, industries are also different. Thus, some techniques used in the IoT application in a community could be compliant, but they could be compliant in the other community. In addition, with the evolution of the IoT, the IoT application will pervasively exist in the people's routine. The change in the people's lives is imperceptible in a short time. But after several years, we believe the change for the society will be huge, and lead to new laws.

Finally, the education where we will tell the users how to securely use the IoT application will be the big challenges. For example, the social engineer attack, such as phishing, is usually successful for users who are not well-trained. The success would lead to disaster for the users' sensitive data, such as identification information, including their privacy data. An effective countermeasure is to educate people how to recognize the fake access points (readers) or servers, which are usually roughly set up and easy to be recognized.

7.2.5 Application Areas and Industrial Deployment

The IoT potential market segments and their current and future applications are briefly summarized in Table 7.1.

7.2.5.1 Global fresh food tracking

In this section, we present an application example on how such a system is deployed for fresh food tracking services. According to our knowledge, approximately 10% of the fresh fruits and vegetables coming from different parts of the world into European market are wasted during transportation, distribution, storage, and retail processes. It causes not only a loss of around 10 billion Euros per year, but also a big threat to the public food safety and carbon dioxide emission. The main causes of fresh food damage during the

Table 7.1. Market segments and the IoT applications.

Market Segments	The IoT Applications
Logistics and Supply Chain Management	Temperature monitoring for consumer goods industry
	Monitoring of hazardous goods and chemicals
	Theft prevention in distribution systems for high value goods
	Container monitoring in global supply chains
	Decentralized control of material flow systems
	Identification of bottlenecks in process
	Supply chain event management
Security Infrastructure	ID Card and passport systems
	e-Token system for Online Authentication
Automation, monitoring, and control of industrial production processes	Industrial automation in general
	Quality control within production process
	Machine Condition Monitoring
	Inventory Tracking and Surveillance
	Monitoring of process parameters like temperature, pressure, flow
Energy & Utility Distribution Industry	Sensor network-based smart grid system
	Automated meter reading
Health care and medical applications at home and in hospital	Patient localization inside large hospital
	Monitoring of vital parameters
	Position and posture monitoring
	Optimization of patient flow in hospital
	Hospital personnel and equipment tracking
	Care for elderly people
	Inventory management
Civil protection and public safety	Monitoring of building integrity for bridges, tunnels, gymnasiums
	Early warning systems for detection of emerging forest fires
	SLEWS — A prototype landslide monitoring and early warning system
	Localization and monitoring of fire fighters and other rescue staff
Learning, Education, and Training (LET)	LET Collaboration application areas
	LET Text based collaboration
	LET Multimedia based collaboration
	LET Learner communication (communication devices managed by the IoT)
	LET Augmented cognition application areas
	LET Privacy application area
	LET Biometric feedback application area
	LET MLR for describing content to be sent over the IoT
	LET Accessibility
	LET Quality processes

(*Continued*)

Table 7.1. (*Continued*).

Market Segments	The IoT Applications
Automation and control of commercial building and smart homes	Building energy conservation system
	Adaptation of living environment to personal requirements
	Monitoring and control of light using occupancy and activity sensors
	Monitoring and control of temperature, humidity, heating, etc.
Automation and control of agriculture processes	Monitoring of growing areas
	Crop disease management
	Nutrient management
	Microclimate control
Intelligent transportation and traffic	Parking management system
	Harbour freight intelligent management system
	Advanced travellers information systems
	Advanced traffic management systems
	Advanced public transportation systems
	Commercial vehicle operation systems
	Advanced vehicle and highway information systems
	Aircraft traffic management systems
	Fleet management systems
	Car-2-car communication for early warning systems
Environment observation, forecasting, and protection	Monitoring of permafrost soil for early detection of problems
	Detection of water pollution in nature reserves
	Temperature monitoring of coral reefs
	Detection of gas leakages in the chemical industry
	Weather observation and reports
	Seismic sensing and flood monitoring
	Environmental pollution including water and air

above handling process are microbial infections, biochemical changes due to biological processes, physical food injuries due to improper environmental conditions, and mechanical damage due to mishandling.

Architecture and Operation Flow of the Global Fresh Food Tracker

Our proposed sensor tags and systems are therefore developed for global fresh food tracking service. As shown in Fig. 7.2, the service is managed by an Operation Center (OC), which controls all the sensor nodes, databases and provides all services to users. Services are accessible through kinds of terminals from a complicated enterprise resource planning (ERP) system to personal laptops and mobile phones. Typical user interface comprises a web based data analysis and visualization tool, a GoogleMap^{TM} compatible route tracking tool,

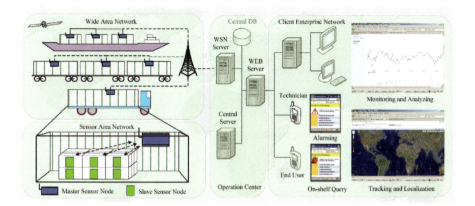

Fig. 7.2 System architecture and operational flow for the fresh food tracker.

and a Short Message Service (SMS) based alarming and query tool for mobile phone users.

The sensor tags (slave nodes) and master nodes (MSN) are deployed based on the two-layer network topology as mentioned in the beginning of the article. The system collects all real-time primary condition parameters, including the GPS coordinate, temperature (T), relative humidity (RH), CO_2/O_2/ethylene concentration, and 3-axis acceleration through mobile and remotely controllable sensor nodes. Furthermore, user friendly service access tools for service registration and specification, real-time data monitoring and tracking, alarming and close-loop controlling, and reliable information sharing are provided as web services based on the service oriental architecture (SOA).

7.2.5.2 Identity security

Although the security and privacy are two important issues in the IoT, an important application is security. Based on the identification technology, a person also can be identified. Some countries start projects to deploy large-scale identity card systems for their citizens. Typically, each Chinese citizen has a second-generation ID card where a chip is embedded and some private enciphered data are stored. The readers are deployed in secure environments, and the reading processing will be operated by a well-trained clerk to ensure the security of the second generation ID card system.

But the public security in the IoT leads to some new and urgent requirements which include:

- The combination of the identities in the IoT: With the rapid development of applications in the IoT, such as e-government, and logistics systems, it is vital for the applications to identify a citizen who is connecting to them. Particularly, there are huge identities of objects in the IoT, and these identities could be linked to one citizen in the real-world. In addition, a citizen usually has many types of digital identities, and each type has lots of digital identities in the IoT. Thus, it is very important and urgent to provide a simple, pervasive and trusted link between citizens and virtual roles in the IoT. After setting up the link, citizens can securely visit the services in the IoT, whereas the government can effectively manage the IoT.
- Authentication for online services: Online-banking and e-government services are more and more popular in the current Internet society. In the services, the first security consideration is to authenticate the user's identity, especially to authenticate whether the user is a person in the real-world. Furthermore, logistics systems are one of the most popular applications in the IoT. Logistics systems are usually the infrastructure of e-business applications. But in the current e-business applications in China, it is hard to authenticate the buyer who is ordering. On the other hand, the buyer also does not want to leak any information to e-business venders and logistics companies. That is, e-business providers and logistics companies should not know the detailed information of the buyer's private identity information. Even logistics companies cannot reason the two different orders for one person. Current digital management systems are hard to meet the security requirements and especially cannot provide an online way to meet it. The current method in China requires a citizen to show his or her ID card, leave a copy, and sign paper files to obtain a digital identity via a face-to-face interview.

The Third Research Institute of Ministry of Public Security in China is developing and deploying a large scale identity management system, referred as eID (electronic id in China), for the identity management and authentication in the IoT. Based on the eID, citizens can easily and securely combine with the virtual roles in the IoT. In the eID system, a citizen can hold a card which supports both contactable and contactless interfaces, and stores the private keys in the card. The card also supports the encryption/decryption algorithms and signature algorithms in the inner chip. Thus, the sensitive data will not be leaked out of the card. As a result, the eID uses the technologies of the IoT to assure the security of the IoT applications. In 2011, eID will be deployed and pre-operated in Shanghai.

Other countries and areas are also developing and deploying similar systems. European Union (EU) is going on a digital identity management system, also referred as eID (EU), among their member states and other allies. The eID (EU) provides a pervasive and cross-border digital identity management service. The goals of eID are similar to eID (EU) except for the IoT. Next, the US government published a strategy report of Identity Ecosystem, which will provide trusted identities in cyberspace. Identity Ecosystem will provide secure, efficient, easy-to-use, and interoperable identity solutions to access online services in a manner that promotes confidence, privacy, choice, and innovation. Furthermore, other countries, e.g. Korean, and companies, e.g. Microsoft and IBM, also set up some digital management systems for online services in the Internet.

7.2.5.3 Smart grid

Smart grid system is an electricity transmission and distribution network which adopts advanced wireless sensor nodes as "ears and eyes" to collect detailed information about the transmission and distribution of electricity. Integrated with robust bi-directional communications and distributed computers, the smart grid is a self-adaptive system to counter fluctuating and unstable demands of electricity in order to improve the efficiency, reliability, and safety of power delivery. Unlike the traditional grid, the smart grid is a digital network keeping pace with the modern digital and information age, allowing a flexible tariff scheduling that encourages clients to use electricity more wisely. Figure 7.3 shows a typical smart grid architecture.

7.2.5.4 Automation, monitoring and control of industrial production processes

In industrial automation, there are numerous tasks to be considered, such as different means of supporting emergency actions, safe operation of the

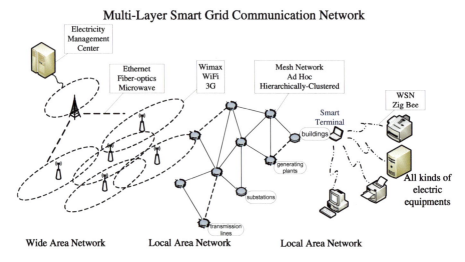

Fig. 7.3 Smart grid.

plant, automated regulatory and supervisory control, open loop control where a human being is part of the loop, alerting and information logging, and information uploading and/or downloading. Some of these tasks are more critical than others. The industrial automation systems are complex and often very expensive. In the future, wireless sensor networks may be applied to realize cost effective and efficient automation with simpler mechanisms, which fulfills the exactly the same functions as the existing problem solutions that have been in use.

Figure 7.4 shows a top view of the industrial wireless network architecture. Wireless HART (Highway Addressable Remote Transducer) and ISA100.11a provide specifications to support wireless process automation applications. The architectural elements are wireless communications systems that consist of a single subnet or multiple subnets connected to a single control room. Otherwise, for a large site with multiple interconnected control rooms, each room can be connected with multiple subnets:

- The subnets are used for control or safety where timeliness of communications is essential;
- The subnets are used for monitoring and asset management;
- The subnets can (but need not) support plant workers wirelessly and also support plant and civil authority's first responders;

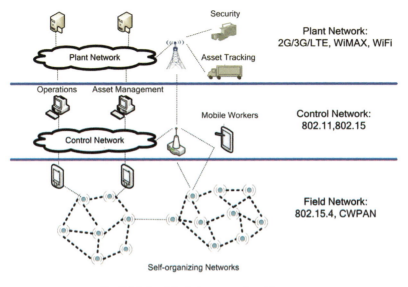

Fig. 7.4 Industrial wireless network architecture.

- The subnets can (but need not) require a proof that a device is authorized to operate in the network;
- The subnets can (but need not) provide limited intrusion resistance within a wireless subnet;
- The subnets can (but need not) provide, within a wireless subnet, a limited higher-layer message confidentiality and resistance to traffic analysis;
- The subnets can (but need not) provide extensive messaging security at the granularity of individual communication sessions.

7.2.5.5 Health care and medical applications at home and in hospitals

Many elderly people must leave their homes to move into a nursing home when the risk for living alone in their own homes becomes too high. For example, people suffering from dementia tend to fall down while fulfilling simple everyday tasks. Sometimes they are unable to stand up on their own and the consequences could be fatal. Researchers from academia and industry are trying to find solutions for this kind of problems involving the elderly, the handicapped, or patients. One of these solutions is to use a special type of sensor network attached to the elderly person or patient. The main advantage of this approach compared to other simpler solution approaches is that the sensor tags are active and smart. The sensor tags can detect uncommon body positions, and the tags can generate and transmit an alarm message when detected. This application can also apply to those who are short-term patients, e.g. recovering from stroke, cancer, major surgeries, and other injuries. The short-term patients can eventually return to their normal lives. The elderly, similar to long-term patients, may be equipped with different sensors in their homes or care centers compared to those who are in short-term care.

The main idea is to attach sensor nodes to the extremities of the elderly person or the patient. The sensor nodes monitor their own spatial orientations and their relative positions to each other. The sensor nodes or tags send the measurement data to a central unit which compares the data with reference information in a database. In case that the measurement data is not acceptable for the person or patient under monitoring, an alarm message is generated and routed to a health monitoring service provider. Another potential solution is the idea of Ambient Assisted Living (AAL), which have sensors built-in a house to monitor the patients, moving toward becoming one of the functions of Smart Homes. The examples discussed in this section are mainly for the home environment but could similarly be used in hospitals.

7.2.5.6 Automation and control of commercial buildings and smart homes: building energy conservation system

Accurate energy consumption monitoring of a buildings electric infrastructure, such as elevator, lighting, air conditioning, fire alarm system, ventilation, high and low voltage power distribution, etc., is one of the key issues to achieve an energy-saving or energy efficient building. In construction of new buildings and updating existing buildings to install the energy consumption monitoring system to conserve energy, the most pressing issues are the high cost of integrated wiring and the high cost of reparations after the update. Therefore, for both the new building and the existing buildings, the best way to transmit a message is through wireless means; however, the traditional wireless systems, such as GSM, WLAN, SCADA, etc., and their power and equipment costs are very high; yet, their network abilities are limited. The Building Energy Conservation (BEC) system based on wireless sensor network technology is considered the best solution for the building energy consumption monitoring as a part of the BEC system.

The BEC system built on wireless sensor network technology collects and distributes information about environment parameters and energy delivery and usage. Wireless sensor network nodes collect the environment parameters, such as temperature. On the other hand, the network nodes are connected with a variety of sensors/devices that collect the energy information on delivery and usage of electricity, water, gas, etc. This sensor networks consisted of hundreds of nodes need to be able to self-organize to provide a reliable wireless network. Energy information can be monitored and processed in real-time for energy stability diagnosis, energy consumption assessment, and energy transformation based on the results of the energy consumption assessment.

7.2.6 Governance and Socio-economic Ecosystems

The rapid growth of the IoT in various applications has evoked much attention of its interoperability, security, and other governance issues. The European Commission has been looking into the needs for the IoT governance for years. According to the European Commission, policymakers should also participate in the development of the IoT alongside the private sector. Stakeholders from governments and industry have formed various organizations to establish the scope, the framework principles and norms for the IoT international governance. Some challenges are indeed policy-related, as highlighted by the World Summit on the Information Society, which encourages the IoT governance designed and exercised in a coherent manner with all the public policy activities related to Internet Governance.

The analysis and status of the current the IoT standards are shown in Tables 7.2 and 7.3.

In the IoT applications, each physical object is accompanied by a rich, globally accessible virtual object that contains both current and historical information on its physical properties, origin, ownership, and sensory context. The incredible amount of information captured by a trillion sensing tags should be well-processed. Therefore, powerful applications are required to transform low-level RFID data into meaningful high-level information. Additionally, the IoT Eco-system requires a secure platform that helps users understand and

Table 7.2. Standards in the IoT applications.

Application Area	Analysis
Air interface standards	These standards are well defined through various different committees as e.g. ISO/IEC SC31, SC17, SC6 and IEEE 802.11, IEEE 802.15, CWPAN and others
Application standards	Application standards suffer under a significant lack of standards
Conformance and performance standards	Conformance and performance standards are beyond the requirements coming from the air interfaces, however, the responsibility lies in the groups developing the air interface standards
Data encoding and protocol standards (often called middleware)	Sufficiently available
Data exchange standards and protocols	Depend on the specific application requirements
Data protection and privacy regulations	Lack of standards. EC is addressing this through ETSI and CEN.
Data standards	Okay
Device interface standards	Okay
Environmental regulations (e.g. WEEE, packaging waste)	Outside the scope of this analysis as this applies for all electronic devices
Frequency regulations	Many frequency bands are globally well regulated. UHF RFID and UWB require a better analysis, where the UHF RFID band Europe is moving closer to Chinese band although using different channel widths. The general global attention for UHF RFID is that high that there is a highly likelihood that all Nations will provide at least one band in the 900–930 MHz area.[1,2]
Health and Safety regulations	Outside the scope of this analysis as this applies for all electronic devices and all RF devices
Internet Standards	Addressed mostly by IETF
Mobile RFID	Defined through ISO/IEC 29143 and NFC standards like ISO/IEC 18092
Real time location standards	Work ongoing and well addressed in ISO/IEC JTC1 SC31
Security standards for data and networks	Lack of standards. EC is addressing this through ETSI and CEN.
Sensor standards	Addressed in ISO/IEC JTC1 SC31 and ISO/IEC JTC1 WG7
The European Harmonisation procedure	Outside the scope of this analysis as this applies for all electronic devices and all RF devices
Wireless Network Communications	Well addressed in IEEE 80

[1] Europe is currently working on releasing of the 915–921 MHz band. In case this is successful, then other countries currently utilizing the same 865–868 MHz band as Europe may follow Europe in using the 915–921 MHz band.
[2] Japan is currently considering a change to the 915–928 MHz band for UHF RFID.

Table 7.3. Standards in the IoT.

Standardization Areas	Other SDOs, Consortia, and Fora
Terminology	ITU-T SG 13, 16, 17; JCA-NID
Requirements Analysis	ISO TC 204, ISO TC 205, TC 211, IEC TC 65, ITU-T SG 13, ITU-T SG 16, ITU-T SG 17, ISA100, IETF 6LoWPAN, ROLL WG, OGC
Reference Architecture	ITU-T SG 13, ITU-T SG 16, ISO TC 204, ISO TC 211
Application Profiles	ZigBee Alliance, OGC, ITU-T SG 5
Sensor Interfaces	IEEE 1451.x, IEC SC 17B, EPCglobal, ISO TC 211, ISO TC 205
Data type and Data Format	ITU-T SG 16, ISO TC 211, ISO TC 205, W3C, IEC TC 57
Communications	IEC SC 65C, IEC TC 57, IEEE 802.15.x; IEEE 1588, IPSO Alliance, ISO TC 205, ISA100
Mobility Support	IETF MANET MIP WG
Network Management	ZigBee Alliance, IETF SNMP WG, ITU-T SG 2, ITU-T SG 16, IEEE 1588
Collaborative Information Processing	OGC; W3C
Information Service Supporting	OGC, W3C, IETF ENUM WG, EPCglobal
Quality of Service (QoS)	ITU-T, IETF
Middleware	ISO TC 205, ITU-T SG 16

control their privacy settings. The IoT Eco-system should have the features of (1) binding of physical objects and virtual objects, (2) real-time location services, (3) timely insights and responses, (4) information security and privacy, (5) information visualization and (6) historical information analysis. There is an example of the IoT Eco-system: the RFID Ecosystem.

The RFID Ecosystem is a scaling, community-oriented research infrastructure creates a microcosm for the IoT at the University of Washington. It is built with the EPC Class-1 Generation-2 RFID tags and readers. It provides the opportunity to investigate applications, systems, and social issues that are likely to emerge in a realistic, day-to-day setting. A suite of user-level, web-based tools and applications for the IoT are developed and deployed in the RFID Ecosystem.

In the IoT Eco-system, each physical object is accompanied by a globally accessible virtual object that contains both current and historical information on its physical properties, origin, ownership, and sensory context. The incredible amount of information captured by a trillion sensing tags should be well processed in security. The features and relevant challenges for IoT Eco-system are listed in Table 7.4.

Table 7.4. Features and challenges for IoT Eco-system.

Feature	Challenges
Binding of physical objects and virtual objects	Small, low-cost, low-power wireless sensing devices
Real-time location services	Robust locating algorithms, especially for passive RFID-based IoT applications
Timely response and intelligent management	Communication protocol, Software platform
Information security and privacy	Cost-effective encryption technique
Information visualization	Data visualization
Historical information analysis	Mass memory, Data mining
Industrial chain integration	Business model

7.3 Technical Challenges of the Internet of Things

To address the challenges of the architectures, we propose:

- View the things as a service is a big challenge of SOA due to performance and cost limitations.
- Automated things composition for the IoT applications.
- Domain control for the IoT applications.
- Cross-domain interoperation and cooperation.

To address the challenges of the network technology, we propose:

- The IoT integration of heterogeneous networks, and system seamless wired or wireless access to various types of networks to cater to various users' communication requirement.
- Device automatic selection of local networks, and adaptation to local communication environments.
- Multiple virtual addresses allocating to devices or objects in the physical world in things to things communication for identification and localization.
- Optimization of devices management, including mobility, network types, communication priority, network handover, and improving the quality and efficiency of the wireless communication system.

To address the challenges of the discovery and search engine technologies, we propose:

- A description language to describe the Things in the IoT. The language must be standardized, scalable, and flexible to vary kinds of

things in different implement environments, such as tags, sensors, back-end servers.
- P2P based discovery and search engine mechanisms and algorithms that take into consideration the issues of sensors (tags) roaming, real-time requirement, privacy protection, massive data, cross-domain interoperation, and different semantics and laws of governance.

To address the challenges of the security and privacy technologies, we propose:

- Light weight ciphers and protocols for sensors (including tags) authentication. In these ciphers and protocols, the performance, energy and cost will be tough in designing, manufacturing and deploying.
- A pervasive, efficient, scalable and robust security service based on cloud computing to support the IoT application. The service should provide the key management, ciphers and protocols evaluation, identity management, and audit.
- Trade of performance, energy and cost with the developing the IoT technologies and application requirement.
- Privacy preservation and anonymity mechanism.
- The behaviour specification of active sensors (including tags).
- Domain- and event-based policy-driven security management.
- Quantified the security level for the application, and provide customized security features.
- Standardization.

To address the challenges of the applications, we propose:

- Discovery of killer applications.
- Integration with the current IT systems.

7.4 Conclusion

The IoT is developing very quickly, and we introduce the technical view to the IoT which includes the architecture models, network and communication technologies, discovery and search engine technologies, security and privacy technologies, applications and technical challenges. We introduce two views, vertical and horizontal, for the Iot architecture models. In the vertical view, the IoT consists of three layers: the sensing layer, the network layer, and the application layer. In the horizontal view, the IoT consists of different domains. Next, we introduce the current network and communication technologies. Then we introduce the P2P-based discovery and search engine technologies, which both will deal with the things roaming and cross-domain cooperation issues. Fourth, we introduce how to assure the security and privacy in the IoT. After the above supporting technologies, we introduce the governance and socio-economic ecosystem in the IoT. Finally, we introduce the potential challenges for developing the IoT.

With the supports of governments and companies in the world, the technologies of the IoT are developing faster than in the past. However these technical challenges also call the researchers, developers and officers to contribute to these on-going efforts to resolve them.

References

[1] "The internet of Things," International Telecommunication Union (ITU) Internet Report 2005.
[2] Wikipedia. Internet of Things, http://en.wikipedia.org/wiki/Internet_of_Things.
[3] A. Malatras, A. Asgari and T. Bauge, "Web enabled wireless sensor networks for facilities management," *Systems Journal, IEEE*, vol. 2, NO. 4, December 2008, pp. 500–512.
[4] US Government, National Strategy for Trusted Identities in Cyberspace, 2010, http://www.dhs.gov/xlibrary/assets/ns_tic.pdf.
[5] EU, Report on the state of pan-European eIDM initiatives, 2009, http://www.enisa.europa.eu/act/it/eid/eidm-report.
[6] R. H. Deng, Y. Li, M. Yung and Y. Zhao, A New Framework for RFID Privacy, ESORICS 2010, pp. 1–18.
[7] P. Peris-Lopez, J. C. Hernandez-Castro, J. M. Estevez-Tapiador, A. Ribagorda, M2AP: A minimalist mutual-authentication protocol for low-cost RFID tags, In: J. Ma, H. Jin, L. T. Yang, J. J.-p. Tsai, (eds.) UIC 2006. LNCS, vol. 4159, pp. 912–923. Springer, Heidelberg (2006).
[8] J, C. Hernandez-Castro, J. M. E. Tapiador, P. Peris-Lopez, J.-J. Quisquater, Cryptanalysis of the SASI ultralightweight RFID authentication protocol, *IEEE Transactions on Dependable and Secure Computing* (2008).
[9] A. Mitrokotsa, M. R. Rieback and A. S. Tanenbaum, Classfying RFID Attacks and Defenses, Information Systems Frontiers Special Issue on RFID, 2009.
[10] A. Juels, RFID Security and Privacy: A Research Survey. Selected Areas in Communications, *IEEE*, 2006.
[11] K. Finkenzeller, RFID Handbook, Fundamentals and Applications in Contactless Smart Cards and Identification, John Wiley and Sons Ltd, 2003.

[12] EPCTM Radio-Frequency Identity Protocols Class-1 Generation-2 UHF RFID Protocol for Communications at 860–960 MHz Version 1.2.0, EPCglobalTM Technical Report, May 2008.
[13] G. P. Joshi and S. W. Kim, "Survey, nomenclature and comparison of reader anti-collision protocols in RFID," *IETE Technical Review*, vol. 25, no. 5, pp. 285–292, September–October 2008.
[14] 800/900 MHz Radio Frequency Identification Application Regulation (Temporarily), Ministry of Informatics Industry, P. R. China.
[15] C. Mutti and C. Floerkemeier, "CDMA-based RFID systems in dense scenarios: Concepts and challenges," *Proc. IEEE Int. Conf. RFID*, pp. 215–222, April 2008.
[16] D. K. Klair, K.-W. Chin and R. Raad, "A survey and tutorial of RFID anti-collision protocols," *IEEE Communications Surveys and Tutorials*, vol. 12, no. 3, pp. 400–421, Third quarter 2010.
[17] E. Fleisch, "What is the internet of things? — An economic perspective," Auto-ID Labs White Paper, WP-BIZAPP-053, Janaury 2010.
[18] E. Welbourne, L. Battle, G. Cole, K. Gould, K. Rector, S. Raymer, M. Balazinska, and G. Borriello, "Building the internet of things using RFID: The RFID ecosystem experience," *IEEE Internet Comput.*, vol. 13, no. 3, pp. 48–55, May 2009.

8

Mobile Devices Enable IoT Evolution From Industrial Applications to Mass Consumer Applications

Prof. Jian Ma[1] and Long Cheng[2]

[1]*Wuxi SensingNet Industrialization Research Institute, China*
[2]*Beijing University of Posts and Telecommunications, China*

8.1 Introduction

It is expected that Internet of Things (IoT) [1] will have profound effect on the efficiency of domain-specific industrial applications, such as intelligent transportation, smart power grid, structure health monitoring, environmental protection, smart agriculture or disaster management. We are witnessing another IT wave following the computers, Internet – IoT is known as the third wave of the information technology revolution. However, IoT is not a mere extension of today's Internet, it involves intelligent end-to-end systems and it covers a wide range of technologies, including sensing, communication, networking, computing, information processing, and intelligent control technologies.

United States, European Union, China and Japan, etc., are investing large resources in exploration on IoT. IBM raised the concept of "Smart Planet" in 2008, which has gained a very high recognition in IoT research field. Subsequently, various countries developed their own IoT strategies in less than a year, e.g., Sensing China, Action Plan for Europe and i-Japan Strategy. China government put a heavy emphasis on IoT research and industrialization. Industrial experts and analytics believe that IoT industry is the next economic growth, its huge market is 10 times the size of the Internet market. According to a market research report, the China market size of wireless sensor network (WSN), core part of IoT at the moment, will grow at an annual compound pace of over 200% in future and reach CNY 20 billion in 2015 [31].

However, the industrialization of IoT is still at the beginning stage, and the current status on IoT in China and other developed countries more is at the research stage. The real IoT applications, at present, are still applied within a small scale. In other words, this is a historic opportunity for China to catch up with the world in the development of IoT.

It is predicted that in 10 years, there will be 50 billion connected devices [19], among which more and more mass consumer devices are connected. This will increase the market demand for applying IoT technologies to mass consumer applications. In this short article, we brief the current and future development of IoT mass consumer applications. The main objective is to share our interest and vision on IoT for the mass consumer market in China.

The characteristic of mass consumer applications is that such kind of applications own immense end users, and thus are closely linked with common people's life. IoT mass consumer applications can increase the awareness of IoT technologies, which can even help to push forward IoT growth in the domain-specific industrial market. Therefore, in order to promote the development of IoT industry, based on China's situation, we also need to tap the full potential of IoT for the mass consumer market.

The remainder of this article is organized as follows. Section 2 introduces the concept of "**IoT mobile device**". In Section 3, we discuss IoT applications in current China mass consumer market. In Section 4, we describe some future IoT mass consumer applications. Finally, conclusions are drawn in Section 5.

8.2 IoT Mobile Device

We are observing that mobile telecommunications enter the new 3G era. Technological advancements in mobile devices such as component miniaturization and decreased energy consumption have enabled them having more features and computing power, e.g., modern mobile devices are often equipped with GPS receiver, camera, acceleration and compass sensors. Mobile devices, especially the smart phones, are more likely to integrate sensing, computing, communication capabilities, and will become very much an extension of our lives. This has made it possible that mobile devices can be used as personal computing platforms and flexible interaction devices with other IoT infrastructures in the physical world.

The concept of "IoT mobile device" first emerged as China's government has made IoT a high priority for national industry [32]. Since mobile devices have become a pervasive part of our everyday lives, people take with mobile devices almost everywhere. The widespread and ubiquitous nature of mobile devices makes them an ideal way for people to interact with the physical world. The basic idea of the so called "IoT mobile device" is to enhance mobile devices with the Radio Frequency Identification (RFID)/Near Field Communication (NFC) technology, so that they either can be identified or are able to read RFID/NFC tags. This can greatly extend the current usage of mobile devices, and thus has a great market prospect.

8.3 IoT Applications in Current Mass Consumer Market

Since August 2009, IoT based applications have quickly emerged in China market. Several factors have contributed to this trend. The immediate impetus was that Chinese Premier Wen Jiabao made a speech in the city of Wuxi indicating that Chinese government would speed up the development of China's IoT industry by introducing a series of support policies [20]. China's consumer electronics manufacturers and service providers have smelled the business opportunities and began to take actions for the IoT mass consumer market.

8.3.1 Mobile Device-centered IoT Mass Consumer Applications

In this section, we discuss some representative IoT applications in current China mass consumer market, including mobile contactless payment, access control, mobile tickets, location-based service and mobile interaction.

8.3.1.1 Mobile contactless payment

Mobile contactless payment is a new and rapidly-adopting alternative payment method. Instead of paying with cash, check or credit cards, a consumer can use a mobile device to pay for a wide range of services and goods. Therefore, the mobile device in this context is referred to as the "mobile wallet." Mobile contactless payment uses RFID/NFC technology to effect payment at subway, vending machines, or point of sale terminals in convenience stores or retailers.

China Mobile, the world's largest mobile operator, has announced plans for a large scale commercial rollout of mobile contactless payments [21].

In fact, the mobile contactless payment is not a new idea, Japan and South Korea have been able to coordinate the complex ecosystem required to extensively deploy a widely used mobile contactless payments system [22]. However, it is reported that China is likely to become the world's largest mobile payments market by 2013. Figure 8.1 shows examples of "IoT mobile devices" for the mobile contactless payment applications in current China market.

8.3.1.2 Access control and mobile tickets

Another application that has a huge potential is to turn mobile devices into access devices, which enables consumers to use mobile devices instead of traditional plastic access cards to verify identity in commercial and consumer settings. Mobile tickets reduce the production and distribution costs connected with traditional paper-based ticketing channels and increase customer convenience by providing new and simple ways to purchase tickets. Figure 8.2 shows examples of extending mobile devices into access control devices or mobile tickets. During the Shanghai World Expo 2010, China Mobile has implanted mobile ticket information into its SIM cards (called NFC-enabled SIM card) [23], so mobile users do not need anything other than a cell phone to gain entry

Fig. 8.1 "IoT mobile devices" for mobile contactless payment in China market.

Fig. 8.2 Examples of extending mobile devices into access control devices or mobile tickets.

to the Expo. User can purchase tickets online; then the tickets information will be downloaded to the SIM cards.

8.3.1.3 Location-based service

A location-based service (LBS) [24] is an information or entertainment service, accessible with mobile device through the wireless communication network and utilizing the geographical position information of the mobile device. LBS is of an IoT application because the spatial temporal location in LBS actually represents the presence of mobile users or elements, thus it connects people or things to the Internet.

LBS can be query-based and provide the end user with useful information (requesting the nearest business or service) such as the nearest Chinese restaurant. LBS can also be push-based and deliver coupons or other marketing information to customers who are in a specific geographical area, e.g., receiving notification of a sale on gas or warning of a traffic jam. Moreover, LBS provides mobile users personalized services tailored to their current location, or navigation to any address.

Juniper Research [25] estimated that the market for LBS will bring in revenues of more than $12.7 billion by 2014. Location-based services are also beginning to gain ground in China, the market has boasted three to four million active users by the end of the third quarter 2010, according to a market report [26].

8.3.1.4 Mobile interaction

The interaction between mobile devices and real-world objects is gaining more attention, because it provides a natural and intuitive way to request services associated with real-world objects [2]. Now more and more mobile devices integrating with cameras, accelerators, compass sensors, RFID or barcode readers have made this direct interaction possible. With the increase of everyday objects augmented with RFID/NFC or visual tags (e.g., 2D barcodes), mobile interaction will be accepted by the mass consumer market. The tag usually has a link to a website address or email address that the mobile devices can access, or it might contain location address and telephone numbers. For example, if the tag contains a web link, mobile user can be navigated from

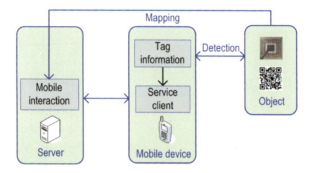

Fig. 8.3 Overview of system architecture for mobile interaction application.

Fig. 8.4 Example of 2D barcode mobile interaction services in China market.

the tags to the web presence of real-world objects. The overview of system architecture for mobile interaction application is shown in Figure 8.3.

In Japan, there is a popular trend using the mobile devices to read information from 2D barcodes, such as on the advertisement posters. Also, there have been the 2D barcode mobile interaction services in China market, as shown in Figure 8.4. A Chinese company also has developed the world's first 2D-barcode decoding chip [3]. As a result, the decoding speed can be increased by more than 10 times compared to software decoding technology.

Although mobile interaction with RFID/NFC tags has been proposed for many years [2], it is mainly performed in the academic study and has not made the jump to mass consumer market, because the RFID tag cost is still high now. However, it can be expected that mobile interaction with RFID/NFC tags is becoming increasingly prevalent as the price of the technology decreases.

8.3.2 IoT Household Appliances

It was Mark Weiser, the former chief scientist at Xerox Parc, first articulated the idea that ordinary household appliances could become embedded with

8.3 IoT Applications in Current Mass Consumer Market 185

Fig. 8.5 IoT household appliances in China market.

information technology, which is his vision of the ubiquitous computing [27]. However, before China launched the national IoT plan, the development of IoT household appliances was still in the concept introduction stage while the market development is limited. Until recently, household appliance manufacturers have intensified efforts to develop the "IoT household appliances", some of them are shown in Figure 8.5.

For example, Haier company applied IoT technologies in the household area and developed the IoT based refrigerator [28]. The IOT refrigerator not only can store food but also can identify information about the food in it and feed back the information to consumers in time by connecting with the Internet, thus achieving the real-time monitoring and management of food. Through the small touch screen on the refrigerator door, consumers are able to check the product information within the supermarket and online orders, and enjoy home delivery services.

IoT based washing machine can automatically adjust energy consumption based on the current electricity grid load and time-of-use price information. It can also automatically identify detergent category, water hardness, washing clothing and weight information, automatically determine and adjust the amount of detergent in the running, select the most appropriate washing procedures and choose the best washing mode.

IoT based air-conditioner can automatically provide the most appropriate and most comfortable temperature, humidity, keep indoor air clean, and also provide early warning and remote monitoring and management services. Users only need to send the order messages through mobile phone to remote control the air conditioner.

Fig. 8.6 Two solutions of IoT smart home in China market.

Furthermore, all the IoT household appliances can be inter-connected and can talk to each other, forming an IoT smart home. Those household appliances will work in a cooperative manner to greatly improve the user experience. Figure 8.6 shows two solutions of IoT smart home in China market. In a smart home, a mobile terminal can be used to control all IoT appliances either locally or remotely, including TV, humidifier, air conditioner, electric curtain, etc.

8.4 Future IoT Applications in Mass Consumer Market

Since the industrialization of IoT is still at the beginning stage, in current mass consumer market, IoT applications are relatively simple and easy to realize. As IoT becomes a reality, there will be more interesting applications brought into the market. In this section, we describe some potential IoT applications in mass consumer market.

8.4.1 Participatory Sensing

The idea of participatory sensing comes from the fact that peer-produced systems can achieve what might be infeasible for stand-alone systems developed by a single entity [4]. A participatory sensing system allows individuals and communities to collect, share and organize information using mobile devices and other mobile platforms. Participatory sensing is also one of the IBM's "Next five in five" Forecast [29]. It forecasts that we all will be data collectors in five years — sensors in devices, cars, wallets and even tweets will collect data that will give scientists a real-time picture of the environment. This data will be leveraged to fight global warming, save endangered species or track invasive plants or animals that threaten ecosystems around the world.

Participatory sensing emphasizes that people participate in the process of sensing and documenting where they live, work, and play [5]. Nowadays, mobile devices are increasingly able to capture images, audio, GPS location, and can interface with the physical environment to allow the measurements of phenomena in real-world, such as public health, traffic monitoring and civic concerns etc. Mobile devices play two important roles in participatory sensing applications, acting as mobile sensors or data collectors.

Actually, the author of this article has proposed mobile enabled Wireless Sensor Networks (mWSN) [30], where the idea is similar as the participatory sensing. That is, mobile devices are exploited to collect and deliver sensory data in mWSN. By exploiting the existing mobile devices to collect sensory data from large scale WSNs has fourfold advantage. Firstly, the data can be delivered to the destination with fewer hops, thus it significantly improves the energy efficiency by reducing the need for multi-hop communication. Secondly, a WSN might become disconnected (or partitioned) into several islands for a variety of reasons, thus exploiting mobile devices can bridge the connectivity gap. Thirdly, mobile users can also access ambient WSNs to get context-aware value added services, e.g., querying nearby sensor networks (i.e., mobile interaction). Finally, there is no extra expenditure of deploying dedicated data collectors. Similarly, the mobile device-based participatory sensing leverages the existing widespread mobile devices, avoiding the need for expensive and dedicated infrastructure.

In recent years, many participatory sensing systems have been proposed, e.g., Common Sense [6], CitySense [7], BikeNet [8]. TrafficSense [9] propose a system that performs rich sensing of road and traffic conditions by piggybacking on smart phones that users carry around with them. The idea of participatory sensing has also been applied in commerce, e.g., MobiShop [10], a participatory sensing system designed to collect, process and share product pricing information from street-side shops to potential buyers through their mobile devices. Besides, it can also serve as an effective advertising medium for retail shops.

Since participatory sensing provides us a very cost effective solution to collect useful sensory data and it is a good match for developing regions, such as India and China. This approach, which can be triggered and used by billions of individuals around the world, is expected to receive more attention in the future.

8.4.2 Augmented Reality

An augmented reality system generates a composite view of the real-world environment for the user. It is a combination of the real scene viewed by the user and a virtual scene generated by the computer that augments the scene with computer-generated sensory input, such as sound or graphics. With the development of more powerful mobile devices and faster wireless broadband networks, it will become inexpensive to put the augmented reality function from the high-end expensive equipments into mobile devices.

The augmented reality could have a major impact on mobile devices in the future IoT mass consumer market. There have been many successful demonstrations. For example, Sekai Camera [11], developed by a Japan company, is an augmented reality iPhone application that tags and overlays information about products and places. It uses the GPS and compass sensors equipped in the mobile device to determine where the device is and at which direction the device is pointing. When user's camera scans a certain tagged item, it shows up on the screen presenting an augmented reality vision of the real world. Sekai Camera combines the ideas of participatory sensing, location-based service and augmented reality. It allows mobile users to share location-based text, photo, voice, and web-links that are posted as the so called air tags. Besides, it also allows mobile users to interact with the floating tags that have been placed virtually in real locations, as shown in Figure 8.7.

Fig. 8.7 Snapshots of Sekai Camera applications.

It is worth mentioning an augmented reality project in China, the Digital Yuanmingyuan [12]. This project is to rebuild the ancient imperial park by digitizing historical material about the park by using the augmented reality technology. An example is shown in Figure 8.8, the earliest layout of the structure (the augmented scene) in Yuanmingyuan is available to the public.

8.4.3 Smart Healthcare

Technological developments of small body-wearable wireless sensors will enable the deployment of Body Sensor Network (BSN) for real time human health monitoring, e.g., blood pressure and heart rate. There has been considerable interest in the development and application of mobile device-centered smart healthcare system, since they offer personalized, unobtrusive monitoring and early detection healthcare services to people. Figure 8.9 shows the overview of system architecture for the mobile device-centered smart

Fig. 8.8 A real scene and the augmented scene of a ruined structure in Yuanmingyuan.

Fig. 8.9 Overview of system architecture for mobile device-centered smart healthcare system.

Fig. 8.10 Preliminary products for the mobile device-centered smart healthcare in China market.

healthcare system. The mobile device collects various physiological changes coming from the body sensor nodes in order to monitor people's health status no matter their locations. The system also allows sending the sensory data as emails or short messages to doctors or medical center, which can help in the early detection, evaluation, and diagnosis.

In China mass consumer market, there have been some preliminary products for the mobile device-centered smart healthcare, as shown in Figure 8.10, where they are characterized by the remote tele-healthcare services. People's health data will be sent to doctors through 3G networks and they can also make a video call with doctors.

8.4.4 Real-world Objects Search

The vision of IoT includes any entities can be connected to the Internet, and publish their output (location, state, etc.) on the Web. This will give real-world objects and places a Web presence that not only contains a static description of these entities, but also their real-time state. The development of IoT is opening up exciting possibilities that in the foreseeable future, people will be able to search real-world entities (i.e., people, places, and things), either through mobile interaction or through the Internet [13], just as the current popular document search service on the Web. Since the state of real-world entities that captured by sensors is highly dynamic, thus, real-world objects searching with a certain state is a challenging problem [14].

Several academic studies have been proposed to support the discovery of real-world entities. MAX [15] is the first work on this area. In Max, RFID tags are attached to real-world objects storing a textual description of these objects. Users can pose queries containing a list of keywords to search for tags holding one or more of these keywords. MAX then returns a ranked list of the matching tags. Snoogle [16] and Microsearch [17] further propose that, sensor nodes that attached to real-world objects carry a detailed textual description of that object in the form of keywords. For example, a node attached to a document binder contains the keyword of reports in this document binder; a node attached to a book would contain the keywords of this book. However, these works require the pre-stored information (metadata) on the tags, there is no mechanism to deal with changing information.

In [13], a real-time search engine for real-world objects is proposed. For example, this search engine could be used to search for rooms in a large building which are currently empty, for bicycle rental stations which have currently bikes available, or for current traffic jams in a city. This system can be combined with the participatory sensing. That is, using mobile device to collect the state information of real-world entities. Actually, this idea has been presented in [18]. The prototype in [18] assumes that objects are tagged with the Bluetooth modules holding the identities of objects (i.e., not a description of the object as in other systems mentioned above). Mobile devices equipped with Bluetooth interfaces are used to detect the presence and identity of such objects. A user can query an object with a certain identity. As a result, the system will return the approximate location of the lost object if it can be found. The overview of system architecture is shown in Figure 8.11.

Although it is still the early days for the real-world objects search applications due to many factors, such as the security and privacy protection issues. However, as increasing numbers of sensors and sensor networks are being

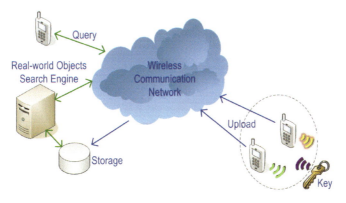

Fig. 8.11 Overview of system architecture for locating everyday items using mobile devices.

connected to the Internet in future, we can foresee that this service will be so helpful in our lives.

8.5 Conclusion

In this article, we addressed the IoT potentials for the mass consumer market. We first introduced the current development of IoT related products in China's mass consumer market, and then discussed some future IoT mass consumer applications. Through reviewing all these applications, we can see that most IoT mass consumer applications in current market are mobile device-centered. Some of them take the mobile devices as the main carrier, enhancing mobile devices with the RFID/NFC technology, so that they either can be identified or are able to read RFID/NFC tags. Other applications use mobile devices as the remote control terminals, e.g., controlling IoT household appliances. As to the future IoT mass consumer applications, mobile devices will play following three important roles: (1) collecting sensory data as mobile sensory data collectors. (2) transmitting sensory data to backend servers; and (3) performing data processing and provide IoT services. We believe that mobile devices enable the IoT evolution from industrial applications to mass consumer applications.

However, we have to admit that although the relevant enabling technologies make the IoT applications described above feasible, there are still many problems and challenges, such as IoT standardization, security and privacy issues, incentive mechanism, business model, etc. To make IoT a reality, significant research efforts need to be conducted. Definitely, we are only scratching the surface of what will be possible in the future IoT mass consumer market. There must be more interesting applications brought into the future market. As IoT becomes a reality, these applications would have a big impact on many aspects of people's life in the coming years.

References

[1] L. Atzori, A. Iera and G. Morabito, "The Internet of Things: A survey", *Comput. Netw.* 54(15), pp. 2787–2805, 2010.
[2] S. Siorpaes, G. Broll, M. Paolucci, E. Rukzio, J. Hamard, M. Wagener, and A. Schmidt, "Mobile interaction with the internet of things", In Embedded Interaction Research Group, 2004.
[3] "Newland releases the world's first 2D-barcode decoding chip", [Online] Available: http://www.newland-id.com/.

[4] A. Kansal, S. Nath, J. Liu and F. Zhao, "Senseweb: An infrastructure for shared sensing", *IEEE MultiMedia*, 14(4), pp. 8–13, 2007.
[5] J. Goldman, K. Shilton, J. Burke, D. Estrin, M. Hansen, N. Ramanathan, S. Reddy, V. Samanta and M. Srivastiva, "Participatory Sensing: A citizen-powered approach to illuminating the patterns that shape our world", *Woodrow Wilson International Center for Scholars*. Septmeber, 2008.
[6] P. Dutta, P. M. Aoki, N. Kumar, A. Mainwaring, C. Myers, W. Willett and A. Woodruff, "Common Sense: participatory urban sensing using a network of handheld air quality monitors", In *Proceedings of the 7th ACM Conference on Embedded Networked Sensor Systems (SenSys '09)*, pp. 349–350, 2009.
[7] R. N. Murty, G. Mainland, I. Rose, A. R. Chowdhury, A. Gosain, J. Bers and M. Welsh, "CitySense: An urban-scale wireless sensor network and testbed", In *IEEE International Conference on Technologies for Homeland Security*, May 2008.
[8] S. B. Eisenman, E. Miluzzo, N. D. Lane, R. A. Peterson, G.-S. Ahn and A. T. Campbell, "The BikeNet mobile sensing system for cyclist experience mapping", In *Proceedings of the 5th international conference on Embedded networked sensor systems (SenSys '07)*, pp. 87–101, 2007.
[9] P. Mohan, V. N. Padmanabhan and R. Ramjee, "Nericell: Rich monitoring of road and traffic conditions using mobile smartphones", In *Proceedings of the 6th ACM conference on Embedded network sensor systems (SenSys '08)*, pp. 323–336, 2008.
[10] S. Sehgal, S. S. Kanhere and C. T. Chou, Mobishop: Using mobile phones for sharing consumer pricing information, Demo Paper in *Proceedings of IEEE DCOSS 2008*, June 2008.
[11] "Sekai Camera", [Online] Available: http://sekaicamera.com/u.
[12] Y. Liu, Y. Wang, Y. Li, J. Lei and L. Lin, "Key issues for AR-based digital reconstruction of Yuanmingyuan garden". Presence: Teleoper. Virtual Environ. 15(3), pp. 336–340, 2006.
[13] O. B. Römer, K. Mattern, F. M. Fahrmair and W. Kellerer, "A real-time search engine for the Web of Things", in *Proceedings of Internet of Things (IOT)*, pp. 1–8, 2010.
[14] K. Römer, B. Ostermaier, F. Mattern, M. Fahrmair and W. Kellerer, "Real-time search for real-world entities: A survey", *Proceedings of the IEEE*, 98(11), pp. 1887–1902, 2010.
[15] K.-K. Yap, V. Srinivasan and M. Motani, BMAX: Human-centric search of the physical world, in *Proceedings of the 3rd ACM conference on Embedded network sensor systems (SenSys '05)*, pp. 166–179, 2005.
[16] H. Wang, C. C. Tan and Q. Li, "Snoogle: A search engine for pervasive environments", *IEEE Trans. Parallel Distrib. Syst.*, 21(8), pp. 1188–1202, August 2010.
[17] C. C. Tan, B. Sheng, H. Wang and Q. Li, "Microsearch: When search engines meet small devices", in *Proc. 6th Int. Conf. Pervasive Comput.*, pp. 93–110, 2008.
[18] C. Frank, P. Bolliger, F. Mattern and W. Kellerer, "The sensor internet at work: Locating everyday items using mobile phones", *Pervasive Mobile Comput.*, 4(3), pp. 421–447, 2008.
[19] [Online] www.ericsson.com/res/docs/whitepapers/wp-50-billions.pdf
[20] [Online] http://english.peopledaily.com.cn/90001/90781/90877/6920367.html
[21] [Online] http://www.nearfieldcommunicationsworld.com/2010/02/11/32670/
[22] [Online] http://en.wikipedia.org/wiki/Contactless_payment
[23] [Online] http://contactlesscities.wordpress.com/2011/03/29/
[24] [Online] http://en.wikipedia.org/wiki/Location-based_service

[25] [Online] http://www.marketresearch.com/
[26] [Online] http://www.geospatialworld.net/
[27] M. Weiser, "The computer for the 21st century", *SIGMOBILE Mob. Comput. Commun. Rev.*, 3(3), pp. 3–11, 1999.
[28] [Online] http://microsoft-news.tmcnet.com/news/2010/01/27/4592428.htm
[29] [Online] http://www.ibm.com/smarterplanet/us/en/ibm_predictions_for_future/overview/index.html?lnk=ibmhpls1/smarterplanet/next_five_in_five
[30] J. Ma, C. Chen and J. P. Salomaa, "mWSN for Large Scale Mobile Sensing", *Signal Processing Systems* 51(2), pp. 195–206, 2008.
[31] [Online] http://callcenterinfo.tmcnet.com/news/2009/12/03/4513662.htm
[32] [Online] http://www.china.com.cn/news/tech/2010-10/03/content_21056852.htm

9

Opportunities, Challenges for Internet of Things Technologies

Dr. José Roberto de Almeida Amazonas

Escola Politécnica of the University of São Paulo, Brasil

9.1 Introduction

This chapter has been written based on a bibliographical review done on December 2nd, 2010, using the IEEE Xplore® search engine. The expression searched was "Internet of Things" and year of publication was restricted to 2010. The search resulted in 150 papers, including conference proceedings and periodicals. Almost all of them were read and provided information to technical sections that will follow.

The papers were initially classified according to the authors' affiliation and subject. Table 9.1 shows the percentage of papers per country and Figure 9.1 the corresponding doughnut chart. It is important to realise that the total number of papers is 160 (>150) and the sum of percentages is larger than 100%. This happens because some papers have more than one author with affiliations from different countries and all affiliations were considered in the classification.

China leads the papers production with an overwhelming 51.3% followed by Europe with 37.3%, USA with 8% and the Rest of World (ROW) with 10%. What do these figures mean? The answer to this question demands great care. The only easy part of the answer is obvious: more than half of the IoT papers written are authored by Chinese researchers. However, this does not mean a leadership in any of the following criteria: quality, originality, technical and/or scientific contribution, worldly knowledge dissemination. The reasons are:

- some of the papers are written in Chinese and then useless for most non-Chinese born researchers;

Table 9.1. Percentage of papers per country.

Country	Number of Papers	% of Papers	Country	Number of Papers	% of Papers
Australia	2	1.3%	Italy	2	1.3%
Bahrain	1	0.7%	Japan	1	0.7%
China	77	51.3%	Korea	3	2.0%
Egypt	1	0.7%	Mexico	2	1.3%
England	4	2.7%	Netherlands	2	1.3%
Finland	1	0.7%	Pakistan	2	1.3%
France	4	2.7%	Poland	1	0.7%
Germany	20	13.3%	Spain	10	6.7%
Greece	2	1.3%	Sweden	1	0.7%
India	3	2.0%	Switzerland	8	5.3%
Ireland	1	0.7%	USA	12	8.0%

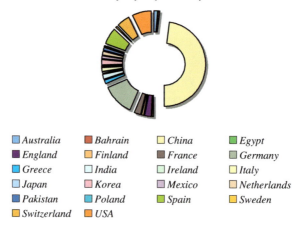

Fig. 9.1 Doughnut chart corresponding to the number of papers per country.

- most of the papers written in English are unreadable, not understandable or simply meaningless. Such papers are generally accepted in "international" conferences hosted in China whose program committee's members work in Chinese institutions;
- there is a small number of papers written in English that are understandable but the technical content is either already known and available in textbooks or corresponds to simple engineering reports that do not advance the state-of-the-art;
- very few papers are high quality ones definitely contributing to advance IoT.

9.1 Introduction   197

Such reality reinforces the need to continuing support the international dialogue and cooperation to leverage Chinese contributions to the level required to promote the development of IoT.

The number of American papers is distorted because IoT-related research and development has been conducted under different names such as pervasive and ubiquitous computing, wireless sensor networks and so forth. More recently, the term used to refer to this area by the NSF in particular is distributed networked sensing systems. Such systems are seen as critical for observing physical, physiological, urban, and social and other such processes at scales, resolution, and timeliness that no other types of instruments can.

European and American papers in general bring important contributions. However, there is a clear difference of approach: most American papers put the technology itself as the main objective while European papers focus on the use of the technology, i.e., they are more user-centric and care about the benefits IoT can provide to the society. North America generally adopts a corporate approach to new technologies. This is either from the perspective of prevailing large corporate entities having a view to project, or from start-ups introducing a disruptive technology or service with the dream of being the corporate of tomorrow (Microsoft, Google, Facebook all started this way). The European Union through its governance structure has defined a set of subjects of exclusive competence that apply across the Union (e.g., customs, competition rules, commercial policy). More significantly for the society are some of the topics of shared competence, particularly: environment, consumer protection, freedom and justice, public health, research, technological development and development cooperation, humanitarian aid.

The ROW papers also provide interesting contributions but without any differentiated characteristic. It is worth to point out that Latin America contributes only with 2 papers from Mexico; there are no contributions from South America. The term IoT is still little used in South America and the developments are concentrated on RFID-enabled private applications. On the other hand, an important Future Internet reference book has been produced under the coordination of South American researchers [1].

Table 9.2 shows the classification of papers according to subject and Figure 9.2 shows the corresponding doughnut chart.

Initially the papers were classified according to the subjects that are dealt with in this chapter, namely: architecture models, network technology,

Table 9.2. Classification of papers according to subject.

Subject	Number of papers	% of papers	Subject	Number of papers	% of papers
Architecture models	27	18.0%	eLearning	4	2.7%
Network technology	17	11.3%	Energy	3	2.0%
Discovery	6	4.0%	Human and social sciences	3	2.0%
Search engines	1	0.7%	Manufacture	2	1.3%
Security	10	6.7%	Recycling	1	0.7%
Privacy	1	0.7%	Health care	3	2.0%
Application areas and industrial deployment	54	36.0%	Smart cities	2	1.3%
Governance	1	0.7%	Agriculture	2	1.3%
Socio-economic ecosystems	17	11.3%	Logistics	3	2.0%
General IoT	7	4.7%	Concept of IoT	3	2.0%
Automotive	1	0.7%	Standards	1	0.7%
AAA	2	1.3%	IoT unrelated	1	0.7%

% of Papers per subject

- Architecture models
- Discovery
- Security
- Application areas and industrial deployment
- Socio-economic ecosystems
- Automotive
- eLearning
- Human and social sciences
- Recycling
- Smart cities
- Logistics
- Standards
- Network technology
- Search engines
- Privacy
- Governance
- General IoT
- AAA
- Energy
- Manufacture
- Health care
- Agriculture
- Concept of IoT
- IoT unrelated

Fig. 9.2 Doughnut chart corresponding to the classification according to subject.

discovery and search engines, security and privacy, application areas and industrial deployment, governance and socio-economic ecosystems. However this initial set had to be extended as can be seen in Table 9.2. As before, the sum of number of papers is more than 150 because some papers cover multiple subjects. It is remarkable the great number of papers concerning applications. This shows that significant effort is being put to transform the IoT into a tangible reality. Most of these papers do not bring any technical or scientific contribution; they could be defined as news about IoT experiments. More important is to realise that the subjects set extension was made to cover the diversity of applications showing the IoT potential to pervade an uncountable sectors of modern life. Last but not least it is worth noting that very few papers were published in governance, privacy and search engines.

Independently of any classification many papers misuse the term Internet of Things. In some cases IoT is just employed as buzzword that helps the paper to be accepted but the content is weakly related to IoT. Other cases show a misunderstanding of what IoT is. The need can be seen for clarification, including criteria on what constitutes IoT applications and services and the foundations for innovation and enterprise.

In considering prospective IoT applications and services questions may be asked to determine their credence:

- What is it that requires access to an Internet structure to fulfil the application or service requirements?
- Can the same requirements be fulfilled without access to an Internet structure?
- What are the implications of using access to an Internet structure to fulfil the requirements, particularly with regard to safety, privacy, security, latency, delay, reliability, speed of response, synchronisation needs, accuracy and temporal relevance of information, system failure, cyber-attack?
- What are the data transfer requirements to fulfil the application and service needs?

Quite often the notion of object-object communication is cited as a distinguishing feature. However, object-object communication can be found in non-Internet based applications, such as can be found in industrial control systems. The first two questions clearly imply that a distinguishing feature of

IoT is the mandatory requirement of accessing an Internet structure to fulfil the application or service. Let's check if IERC and CASAGRAS proposed IoT definitions help to answer such questions.

IERC Definition: *The Internet of Things is an integrated part of the Future Internet and could be defined as a dynamic global network infrastructure with self-configuring capabilities based on standard and interoperable communication protocols where physical and virtual "things" have identities, physical attributes, virtual personalities and use intelligent interfaces, and are seamlessly integrated into the information network.*

In the IoT, "things" are expected to become active participants in business, information and social processes where they are enabled to interact and communicate among themselves and with the environment by exchanging data and information "sensed" about the environment, while reacting autonomously to the "real/physical world" events and influencing it by running processes that trigger actions and create services with or without direct human intervention.

Interfaces in the form of services facilitate interactions with these "smart things" over the Internet, query and change their state and any information associated with them, taking into account security and privacy issues.

CASAGRAS Definition: *A global network infrastructure, linking physical and virtual objects through the exploitation of data capture and communication capabilities. This infrastructure includes existing and evolving Internet and network developments. It will offer specific object-identification, sensor and connection capability as the basis for the development of independent federated services and applications. These will be characterised by a high degree of autonomous data capture, transfer event network connectivity and interoperability.*

Both definitions mention the Internet but they allow an interpretation that the access to the Internet is an optional feature of an IoT application. In the IERC definition they are "seamlessly integrated into the *information network.*" In the CASAGRAS definition "This infrastructure *includes* existing and evolving Internet ...." "*Information network*" could be a LAN and "*includes...*" does not mean that it has to be used.

On the other hand both definitions impose that objects have sensory capability. This is an arguable requirement. If an object is "sensed," the data is used to make a context-aware access to a data server through the Internet, the retrieved information promotes an action either on the object or the environment, this is a legitimate IoT application/service.

Assuming that "things" and "objects" can be used interchangeably the IERC and CASAGRAS definitions make to different assignments:

IERC- "thing" ← (identity, physical attributes, virtual personalities);
CASAGRAS- "object" ← (identification).

In this case there is a clear need to pursue a work of harmonisation of definitions and minimisation of vagueness.

IERC definition states that *Interfaces in the form of services facilitate interactions with these "smart things" over the Internet...* favouring SOA (Service Oriented Architectures) and Web Services approaches. Albeit being well accepted choices they are also limiting and have performance problems.

It cannot be said that the misuse of the term IoT in many papers derives from less than precise definitions. These definitions are not cited by any paper. There is no reference such "according to IERC/CASAGRAS definition of IoT this work reports a new service ...." They may be unknown and/or not accepted. This fact implies the need of a continuing and evolving effort of international dialogue and dissemination activities to arrive at a common understanding of what IoT is.

After this introduction this chapter will discuss the selected subjects and will conclude providing a contribution to an IoT global vision from the perspective of South America.

9.2 Architecture Models

Figure 9.3 shows the architecture models proposed by ITU [2] and EPCglobal [3]. The ITU's architecture has the merit of being a pictorial representation of the IoT definition emphasizing the distinguishing feature of accessing an Internet structure (Next Generation Network — NGN) to fulfil the application or service. Such model does not pave the way towards implementation. It is clearly insufficient and motivates the proposal of new models. The EPCglobal's architecture is not an IoT model. It was create having in mind the implementation of RFID enabled AIDC (Automatic Identification Data Capture) applications. As it is drawn it has the merits of informing the status of the standardisation process. As it does not cover the IoT features it also acts as a driver for new models proposals.

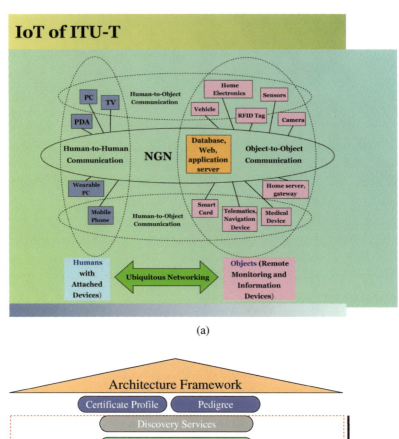

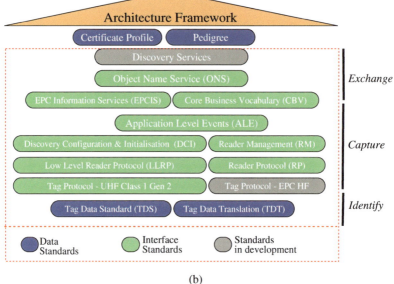

Fig. 9.3 (a) ITU architecture model; (b) EPC global architecture model.

9.2 Architecture Models

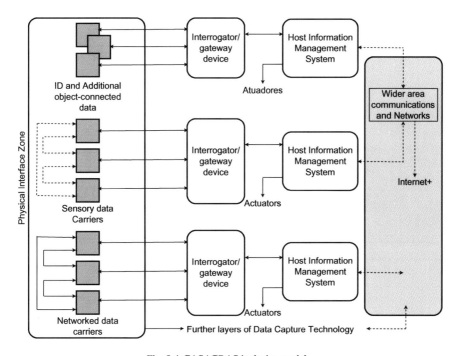

Fig. 9.4 CASAGRAS inclusive model.

The CASAGRAS initiative recognizes the aforementioned insufficiencies and proposes its inclusive-model depicted in Figure 9.4 [4]. This model has been introduced as the reference framework for the CASAGRAS considerations, is more demanding in its outlook and realisation than a number of the restricted technology proposals, it is a vision that can be approached in a staged, standards-supported manner. The framework considerations may be grouped into those that relate to the various layers distinguishable between the real world objects and the integration with the evolving Internet. This model, however, may be misleading as it is possible to consider the following signal path: identification data are captured and sent to the interrogator; following the resolution process the information is sent to the host information management system that may directly drive the actuators *without accessing the Internet*.

Certainly that is not the intention of the proposed model but reinforces the need to a continuing effort to arrive at the definitive model.

In Japan it has been proposed the uID architecture depicted in Figure 9.5 [5]. This architecture consists of ucode tags such as RFID, bar codes, RF and IR

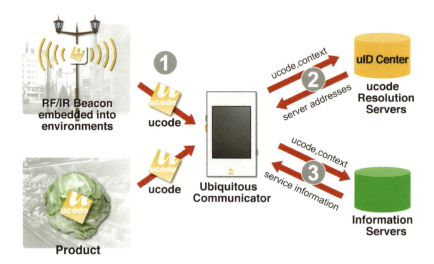

Fig. 9.5 uID architecture.

beacons, user terminals, a ucode resolution server, and application information servers. The uID architecture is purely network based, so ucode tags contain only a ucode, and the ucode resolution and application information servers contain all the information about objects and places. By separating the ID and information, users can easily acquire the latest information on an entity, update that information, and obtain information on other entities related to that entity. On the one hand this architecture has the merit of being purely network based and thus implementing the distinguishing feature of IoT. On the other hand it may present scalability problems when the number of Internet-connected objects reaches the order of many $10^6 \approx 10^9$ devices. The generated traffic to access the resolution and application information servers will be huge and the latency may become unacceptable.

The bibliographic review indicates that many other architectures are being proposed to fulfil the needs of new applications and services.

Akribopoulos et al. [6] present a concrete framework, based on web services-oriented architecture, for integrating small programmable objects in the *Web of Things*. Functionality and data gathered by the Small Programmable Objects (SPO) are exposed using Web Services. Based on this, by exploiting XML encoding, SPO can be understandable by any web application. The architecture proposed is focused in providing secure and

9.2 Architecture Models

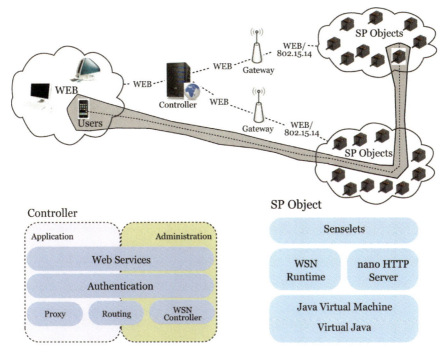

Fig. 9.6 Web-Services-oriented architecture.

efficient interoperability between SPO and the web. Figure 9.6 illustrates the architecture.

SPOs are in the lowest layer of the system. Each wireless sensor device is running a software component. Protocols for the discovery and the communication with the infrastructure and other sensors are provided. It is also implemented a nano HTTP server, which handles each HTTP request in a specific session. After the parsing of a HTTP request the corresponding Senselet is executed. Each Senselet is a tiny Web Service which implements a specific function.

The **Controller layer** (see Figure 9.6) lays upon the controller service and the gateways of SPO sub-networks. This layer provides most of the functionality acting as a core middle-layer in the architecture and is exposed through a Web Services API.

The **Client layer** is the topmost layer and offers two different components allowing users to control the overall system. One is the Web Site, which connects to the proxy, acquires all the necessary data and offers to the user a complete interface to monitor and control the overall system. The second one is the standalone application, which additionally provides direct communication with the sensors omitting the intermediate layers of proxy and gateways. A SPO can act as a client too.

It is important to point out that this is not an IoT architecture, properly named by the authors as Web of Things architecture, that may be considered as a subset of the IoT in which **the actuation is limited to exposing the object to the user**.

Lu Tan et al. [7] depart from the CASAGRAS inclusive model and propose the 5 layers architecture model shown in the left side of Figure 9.7 to implement it.

The authors claim that connecting every object and make them communicate independently is a very attractive vision. It is reasonable to imagine many cases in the future that a thing needs to "talk" to another thing, but is it really necessary that an object "talks" to all other objects? In fact, the

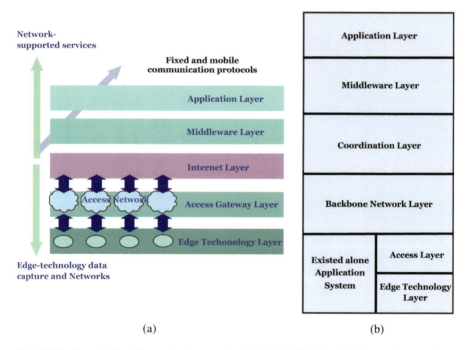

Fig. 9.7 Five layers IoT architecture implementation of the CASAGRAS model; (b) Introduction of the coordination layer.

main connections of an object are with those objects which are in the same IoT application system as it is. So, the IoT is made up of many different IoT application systems. Presently many application are referred as IoT applications as, for example, EPC Global, smart hospital and so on which seem to work well. However these application systems work alone, and even though an object communicates mainly with objects in the same application system, in the future the communication between different application systems will foster their collaboration. The lack of global standards may lead to interoperability problems. Only if the interoperability problems are solved it will be possible to have a real IoT. The authors propose to solve the problem by adding a **Coordination Layer** into the IoT architecture as can be seen in the right side of Figure 9.7. The **Coordination Layer** processes the structure of packages from different application systems and reassembles them to an unified structure which can be identified and processed by every application system.

The authors have identified an important problem: heterogeneity and lack of global standards lead to interoperability problems, but **the proposed Coordination Layer does not seem to solve the problem** because the *unified structure which can be identified and processed by every application system* is simply a **different way of saying that the solution is an accepted global standard**. This will not happen in a foreseeable future: heterogeneity will not disappear from one day to the other.

Kortuem et al. [8] are working toward an alternative architectural model for the Internet of Things as a loosely coupled, decentralized system of smart objects — that is, autonomous physical/digital objects augmented with sensing, processing, and network capabilities. In contrast to RFID tags, smart objects carry chunks of application logic that let them make sense of their local situation and interact with human users. They sense, log, and interpret what's occurring within themselves and the world, act on their own, intercommunicate with each other, and exchange information with people. The vision of an Internet of Things built from smart objects raises several important research questions in terms of system architecture, design and development, and human involvement. For example, what is the right balance for the distribution of functionality between smart objects and the supporting infrastructure? How do we model and represent smart objects' intelligence? What are appropriate programming models? And how can people make sense of and interact with smart physical objects? Three canonical smart-object types (see Figure 9.8)

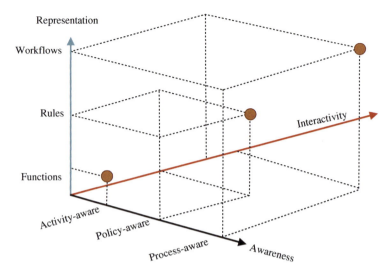

Fig. 9.8 Smart-object dimensions.

have been identified that represent fundamental design and architectural principles: activity-aware objects, policy-aware objects, and process-aware objects. These types represent specific combinations of three design dimensions:

- Awareness is a smart object's ability to understand (that is, sense, interpret, and react to) events and human activities occurring in the physical world.
- Representation refers to a smart object's application and programming model — in particular, programming abstractions.
- Interaction denotes the object's ability to converse with the user in terms of input, output, control, and feedback.

Table 9.3 summarizes these object types and how they relate to the three design dimensions just introduced.

Smart objects' true power arises when multiple objects cooperate to link their respective capabilities. Consider an example of cooperating smart objects, as the safety-aware chemical drum. It is a policy-aware smart object whose application model consists of a set of rules for determining to what extent workers handle it in accordance with safety rules. When we bring multiple smart drums together in close physical proximity, they act as a collective system: drums let each other access their respective rule sets and can thus

9.2 Architecture Models

Table 9.3. Summary of smart-object types.

	Awareness	Representation	Interaction	Augmentation	Example application
Activity-aware object	Activities and usage	Aggregation function	None	Time, state (on/off), vibration	Pay-per-use
Policy-aware object	Domain-specific policies	Rules	Accumulated historical data, threshold warnings	Time, vibration, state, proximity	Health and safety
Process-aware object	Work processes (that is, sequence and timing of activities and events)	Context-driven workflow model	Context-aware task guidance and alerts	Time, location, proximity, vibration, state	Active work guidance

make collective assessments about their safety status as a group (for example, whether the overall volume of all drums exceeds a dangerous limit). In this example, the drums achieve cooperation via a peer-to-peer (P2P) reasoning algorithm for collocated smart objects, in which the reasoning process physically "jumps" from one smart object to the next. All drums that have been part of the collective assessment display notices for users.

It is important to realise that **the aforementioned example as described is not a real IoT application because the set of rules can be stored in a local server and there is no mandatory access to the Internet.**

This article also emphasizes that smart objects provide a distributed architectural model for the Internet of Things. Due to their dual nature as physical and digital entities, such objects highlight the fact that the **IoT can't be viewed only as a technical system but must also be considered as a human-centred interactive one.** This implies that **it is necessary to expand smart-object design beyond hardware and software to include interaction design as well as social aspects.**

Guinard [9] positions the Web of Things as a refinement of the Internet of Things by integrating smart things not only to the Internet (i.e., to the network), but also to the Web (i.e., to the application layer). The main goal is to explore the architecture and tools necessary to build a distributed architecture for smart things which fosters serendipitous re-use of smart things to create opportunistic applications. Just as people create Web mashups involving Web and Web 2.0 services, they should be able to create "physical mashups" mixing services from the real and virtual worlds together. To achieve this goal, the approach is to use patterns commonly used on the Web. First, by embedding Web servers on smart things and adapting the REST (REpresentational State Transfer) architectural style to the physical world. The essence of REST is to focus on creating loosely coupled services on the Web so that they can be easily reused. REST is actually core to the Web and uses URIs (Uniform Resource Identifiers) for encapsulating and identifying services on the Web. In its Web implementation it also uses HTTP as a true application protocol. It finally decouples services from their presentation and provides mechanisms for clients to select the best possible formats. As a consequence of the proposed architecture, smart things and their functionality get transportable URIs that one can exchange, reference on Websites and bookmark. Things are then also linked together enabling discovery simply by browsing. The interaction with smart things can also almost entirely happen from the browser, a tool that is ubiquitously available and that most of the people understand well. Furthermore, smart things can then directly benefit from the mechanisms that made the Web scalable and successful such as caching, load-balancing, indexing and searching.

Although an interesting idea **it seems to suffer from the scalability problem that arises when we consider the number of objects that will be connected to the Internet**. The idea relies upon well known and robust technology but requires further development to be employed in large scale.

Guinard et al. [10] explore the integration of real-world and enterprise services. The protocols usually used do not offer uniform interfaces across the application space and are too complex to integrate with traditional enterprise applications. To ensure interoperability across all systems, recent work has focused on applying the concept of SOA, in particular web service standards (SOAP, WSDL, etc.) directly on devices. A set of requirements to facilitate the querying and discovery of real-world services from enterprise applications is introduced:

- **R1: Minimal Service Overhead.** As most real-world services are offered by embedded devices with (very) limited computing capabilities there is a need for a lightweight service-oriented

paradigm which does not generate too much overhead compared to using functionality through the proprietary APIs.
- **R2: Minimal Registration Effort.** A device should be able to advertise its services to an open registry using network discovery. The process should be "plug and play", without requiring human intervention. A device should also be expected to provide only a small amount of information when registering.
- **R3: Support for Dynamic and Contextual Search.** It should be possible to use external sources of information to better formulate queries. Furthermore, the queries should go beyond simple keyword search and take into account user-quality parameters such as context (e.g., location, Quality of Service (QoS), application context). Support for context is essential as the functionality of most real-world devices is task-specific within a well-defined context (e.g., a building, a manufacturing plant, etc.).
- **R4: Support for On-Demand Provisioning.** Services on embedded devices offer rather atomic operations, such as obtaining data from a temperature sensor. Thus, while the WSN platforms are rather heterogeneous, the services that the sensor nodes can offer share significant similarities and could be (re)deployed on-demand per developer request.

The authors' key contribution is the service discovery process for real-world services called Real-World Service Discovery and Provisioning Process (RSDPP), which goal is to assist the developers at development time in the discovery of real-world services to be included in composite applications. Along with increasingly dynamic infrastructures where mobile devices appear or disappear from the network at operation time, there is a strong need for tools to simplify the management and interconnection of networked devices. Network discovery is a central process in ubiquitous and distributed computing. In contrast to the user-oriented discovery, network discovery enables machines to automatically register themselves and advertise their services on the network. In a way, network discovery is the bootstrap of service discovery for end-users. In line with the previous but bringing a critical view, Nain et al. [11] discuss the growing interest in leveraging SOA, with execution platforms such as OSGi (Open Services Gateway initiative) that make it possible to smoothly integrate the IoT with the Internet of Services (IoS). The paradigm of the IoS indeed offers interesting capabilities in terms of dynamicity and interoperability. However typical SOAs fail to provide the loose coupling and proper separation between types and instances that are needed in domains that involve "things" (e.g., home automation). For instance two light appliances may offer the same type of service (turning light on and off) but different actual services, if only because they are located in different rooms. These loose coupling and proper separation between types and instances are however well known in Component Based Software Engineering (CBSE) approaches. The requirements an execution environment fully adapted for IoT and IoS integration must comply to are:

- **Rq1: An explicit and independent reflexive model of the architecture living at runtime.** Reflecting the actual application, the model makes it possible to reason about the application state. Then an adaptation engine is able to select, test and validate an adaptation scenario on the model, before actually performing the adaptation on the running system. Component-based execution systems often offer introspection capabilities making it possible to build a model view of the running system. SOA execution platforms do not have such an ability.
- **Rq2: Components coupling managed from outside.** For the components to be highly independent, they must not embed any dependency resolution mechanism. Moreover, this extraction would make it possible to modify the resolution policies, or change the connections to adapt the system with no need to deal with business components. Having a clear and explicit description of the relations between components gives a better understanding and makes the analysis of the system much more accurate and so, leads to better adaptation decisions. The component connections are often explicit in Component Based systems, but are never in SOA, and dependencies resolutions are even hard coded in services.
- **Rq3: Interoperability and opening to the outside world** is an essential principle in IoS. The goal is to offer a service in a standardised way to any other system that would like to use it. Even if the system is managed as a component based application, any

third party application must be able to use the services offered by the managed devices (IoT speaking) and more generally components. Services must thus be exposed as classical services, while their "component like" management should remain hidden. This is natively offered by SOA using interfaces and registries to exposes services to the world. Component-based applications are in the other hand, living in their close world.

- **Rq4: Hot deployment ability** is absolutely necessary to ensure future evolutions and adaptations to the protocols and devices. The execution platform must support dynamic deployments and adaptations of the application during runtime with no restart. SOA considers that services appear and disappear at any time. The hot deployment is thus essential and natively taken into account. Component-based applications do not address this concern.
- **Rq5: Minimize the adaptation time.** Another strong constraint working with "things" is that the entire time of realisation of an action starting from the moment a person acts on a sensor, to the moment something happens, must be less than 250 milliseconds for it to be considered as immediate by a human person. The reconfigurations of the system must fit within this constraint, or more specifically, the transition time from a stable configuration to another one should not exceed this limit.

The discussion presented in this section clearly shows that there are too many IoT architectures proposed and none of them is the definitive one. This is consistent with the IoT-A [12] (Internet of Things — Architecture) project statements about the current status:

- Fragmented architectures, no coherent unifying concepts, solutions exist only for application silos;
- No holistic approach to implement the IoT has been proposed yet;
- Many island solutions do exist (RFID, Sensor nets, etc.);
- Little cross-sectorial re-use of technology and exchange of knowledge;
- In essence, there are only Intranets of Things existing.

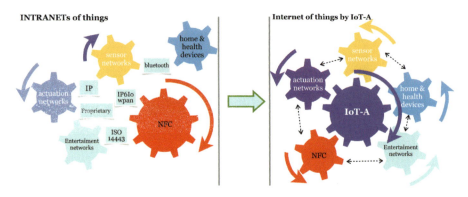

Fig. 9.9 Internet of Things by IoT-A.

The IoT-A project proposes the creation of an architectural reference model for the IoT as well as the definition of a set of key building blocks to lay the foundation for a ubiquitous IoT. Its view is summarized in Figure 9.9.

Besides the ongoing efforts sponsored by the European Commission in the realm of the Framework Programme 7 represented by CASAGRAS2, IoT-i, IoT-A projects, under the supervision and coordination of the IERC, **there is still a clear need to increase such efforts to arrive at an all-inclusive and evolving reference model.**

The universe of the Internet of Things encompasses several elements such as technologies, business processes, standards etc. that are represented by reference models and architectures, in an attempt to understand the use that society is making of them or to propose new solutions for society's demands.

By nature a very fast evolving area, the IoT presents ever new challenges to be faced and constantly demands innovative ways and techniques to be created, merely to represent its very complex universe.

For example, the approaches taken by the Socio-technical Systems Engineering and Model Views propose to put the human being inside the system, promoting user-centric modelling. This is not enough, as we need now to include the environment into the model too, due to the evolution of the wireless technologies, the advent of wireless sensor networks and the emergence of Internet-connected Objects (IcO).

One could envisage to organize part of the aspects and characteristics of the IcO[1] universe into planes. The first plane is the IcO **Technological**

---

[1] In this text, Internet of Things (IoT) and Internet-connected Objects (IcO) are used interchangeably.

Plane, which in an up-to-date approach must also consider the Users and the Environment. The other planes are: **Security**, **Business Processes**, **Integrated Management and Control**, **Regulations**, and **Human Factors**.

These planes represent the existing IcO dimensions that drive, influence and impact the development of services because the interaction between them results in restrictions, enforcements, coordination and challenges to be overcome. It is necessary to achieve a better understanding of this interplay to develop a methodology for integrated roadmap building.

9.3 Network Technology

9.3.1 A Lean Transport Protocol

According to Glombitza et al. [13] the seamless and flexible integration of IoT devices ranging from simple sensor nodes to large-scale Enterprise IT servers are the basis for novel applications and business processes not possible before. Wireless Sensor Networks (WSNs) are no longer limited to simple "sense and send" applications. Now, they can trigger business processes or actions can be triggered on them by a business process being executed on the Internet. Currently, despite the amount of research targeting middleware solutions for sensor networks and resource constrained devices in general, WSNs are not widely used in industry and are not well integrated into Enterprise IT systems. This is mainly caused by the different characteristics and demands of Enterprise IT systems and WSNs. In Enterprise IT systems, it is very important to adapt applications and business processes quickly and flexibly in order to react to changing market requirements. Furthermore, the inter-system compatibility of the used middleware technologies is more important than their performance. Today, virtually all Enterprise IT systems are realized as SOAs, which are implemented using standard Internet technologies, such as XML, Web Services, and the Business Process Execution Language (BPEL). WSNs on the other hand are optimized in terms of low energy consumption as well as low cost and hence CPU power, memory, and bandwidth are severely limited. Hence, these technologies are not applicable in WSNs out-of-the-box.

A key challenge is the efficient message exchange between WSN nodes and Internet nodes. The authors propose a new Web Service transport binding and a novel transport protocol for resource constrained devices called LTP (Lean Transport Protocol). The protocol is not limited to exchanging messages

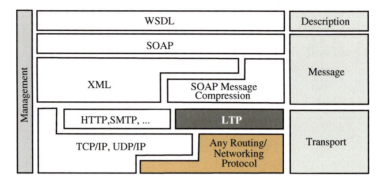

Fig. 9.10 Web Service technology stack (the new LTP protocol is represented in black).

Table 9.4. URL of an LTP Web Service endpoint.

Protocol	RCD	Host/NoRCD	Port	Web Service
ltp	123	host.itm.uni-luebeck.de	555	Web Service
ltp://123@host.itm.uni-luebeck.de:555/Web Service				

between different network types such between the Internet and RCNs (Network of Resource Constrained Devices) but can also be used within a single network. Figure 9.10 depicts the Web Service technology stack and the new LTP protocol is represented in black.

LTP's components are: (i) A network agnostic addressing scheme, (ii) a message transport for UDP/IP, TCP/IP, and resource-constraint environments such as WSNs, (iii) support for arbitrary XML serialization/compression schemes, and (iv) an efficient packet serialization and fragmentation layer.

LTP is providing a comprehensive addressing schema (Platform Independent Addressing Schema) allowing to transparently address Web Service endpoints in RCNs as well as in Enterprise IT systems. LTP endpoint addresses are following the regular URL (Uniform Resource Locator). Table 9.4 illustrates a URL of an LTP Web Service endpoint.

The evaluation shows that LTP achieves a compression ratio of 97.4% compared to a standard WS-addressing packet in conjunction with an extremely low processing overhead compared to XML-based variants (only 0.95% on a PC and 14.6% on a sensor node). The implementation of LTP runs on a variety of hardware platforms and only consumes 7 kB of static code memory and a maximum of 1.5 kB of RAM memory. LTP will be used as enabling technology to develop self-describing Web Services on RCDs as well as improving the integration of RCDs into business processes.

9.3.2 A Multi-path Routing Algorithm

With the advance in sensing technologies and the strong support from governments, industries, and academia, we can envision a world in the near future with sensors deployed at various locations, i.e., surveillance cameras in roads, subways and airports for monitoring traffic conditions, suspicious activities and sensing environmental parameters. To provide an effective connectivity between the remote video cameras and the central location, it is necessary develop wireless and wired hybrid networks that fulfil the QoS requirements. Due to the strict requirements of real-time video applications such as bandwidth guarantees and delay constraints, multi-path routing comes to the fore since the bandwidth is limited in the actual networks. At the same time, in wireless and wired hybrid environments, the packet loss ratio of the links, especially the wireless links, may undermine the performance of multi-path routing considerably. Wang et al. [14] investigate the algorithms for the reliable multi-path routing to find a set of source to destination paths such for which the maximum delay is minimized while the reliable aggregated bandwidth satisfies the bandwidth requirement under the coexistence of packet loss. This problem is NP-hard, and the state of the art algorithms do not consider the packet loss ratio. The authors developed a maximum flow-based heuristic algorithm as the benchmark, and then proposed a fully polynomial time approximation-algorithm that can compute a $(1 + \varepsilon)$-approximation solution. Numerical results based on both well-known Internet topologies and large size random networks demonstrate the effectiveness of the proposed approximation algorithm.

9.3.3 Grey Decision-making in the IoT

Figure 9.11 shows the hierarchical IoT infrastructure model proposed by Liu et al. [15]. In an IoT infrastructure system three major operating components are necessary including RFID device manipulation, network interaction and database management. Network interaction consists of basic services also named APIs provided by development platforms.

A network switch is selected if the corresponding network's QoS factor excels a given threshold. The cost of using a network n at a certain time

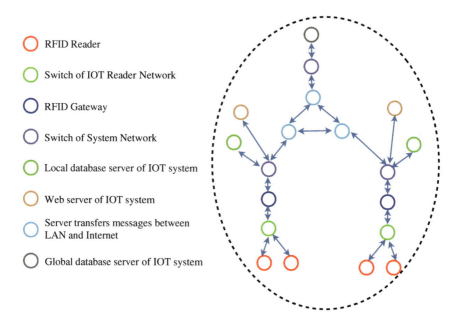

Fig. 9.11 Hierarchical IoT infrastructure model.

is a function of several parameters: the bandwidth it can offer B_n, the power consumption of using the network access device E_n, and the cost $C_n = C_{\text{User}} + C_{\text{Link}}$ of this network.

From the point of view of an application, the nodes and the communication traffic in IoT can be considered as the transmission of a datagram that involves multiple paths routing (see Figure 9.11.) for several pairs of origin and destination nodes. Because of the increase of communication traffic, computing complexity and mobility in an IoT environment, many tasks cannot be finished by a single device. So several devices have to cooperate to finish the target task. How to choose a suitable service from all the useable services regardless of user's and object's location are most important steps in the IoT domain. A possible approach to address the problem is to use a Grey decision-making algorithm. In case of conflicting alternatives, a decision-maker must consider imprecise or ambiguous data, which is norm in this type of decision-making problems. A grey number is a number whose exact value is unknown, but a range within which the value lies is known. The theory of grey systems consists of the following main concepts and results: (i) foundation, consisting of grey numbers, grey elements and grey relations; (ii) grey systems analysis, including grey incidence analysis, grey statistics, grey clustering, etc.; (iii) grey systems modelling, through the use of generation of grey numbers or function so that hidden patterns can be found; (iv) grey prediction; (v) grey decision-making; (vi) grey control. The GRA (Grey Relational Analysis) has some advantages: (i) It involves simple calculations and requires a smaller number of samples; a typical distribution of samples is not needed. (ii) The quantified outcomes from the grey relational grade do not result in contradictory conclusions to qualitative analysis. (iii) The grey relational grade model is a transfer functional model that is effective in dealing with discrete data. The method has been evaluated by simulation and proved its usefulness for assigning resources in an IoT environment.

9.3.4 Network Evolution

According to Rouzic [16] IoT is presently characterised by ≈ 3 connected devices per household and most applications are concentrated on the logistics arena. There is a trend to applications extend to cover surveillance, security, healthcare, documents tracking and tele-operation fields. It is foreseen a multiplication of connected devices reaching more that ten times today's number of devices. Machine 2 Machine (M2M) communication will become common place. Among the challenges it is worth noting the management of customers' devices and the security of critical applications. IoT will have a strong impact on the generated volume of traffic and its bursty nature. It is questionable if the prevalent LRD (Long Range Dependence) Internet traffic model will be valid for IoT.

A major trend is to consider network as a service, i.e., the operator's network as a saleable resource and a major requirement is the ability to deliver on-demand network services. On the one hand the operator will provide network bandwidth on demand and on the other hand it needs to improve its billing capabilities. In this scenario the major challenges to be faced are SLA (Service Level Agreement), QoS (Quality of Service) and QoE (Quality of Experience) management. In addition, addressing and mobility management also represent difficult issues to be solved. Some of the most important technology enablers are: network virtualisation, automation and distributed (cloud) computing. Altogether there will be a great impact on the flexibility requirements on the planning stage for highly changing traffic distribution and volume (certainly very difficult to predict) demanding new CAC (Connection Admission Control) algorithms to be able to accept traffic from any-where at any time and be able to price it.

9.3.5 Comments

Three of the works summarized above deal with very specific problems: a network transport protocol, a multi-path routing algorithm and a Grey-based resources assignment algorithm. The last one presents an overview of the network evolution and the challenges to be faced. It can be seen that the problems to be solved in the IoT environment demand new developments to tackle the new level of complexity. Network virtualisation, multi-path multi-constraint QoS-aware routing, traffic modelling and estimation are some of the technological pillars.

The bibliographical review resulted in 17 articles classified as "network technology." Six of them have also been classified as "architecture models." It is remarkable to realise that from the 11 "network technology only" papers only 1 has been written by a non-Chinese researcher. What is the meaning of this? Are the researchers from the EU not interested or concerned about IoT network technology issues? A possible answer perhaps may be found in the FP7 Work Programme 2011 [17].

On the one hand, IoT is the object of **Objective ICT-2011.1.3 Internet-connected objects:** *The objective is to provide the* **architecture** *and technological foundations for developing context-aware, reliable, energy-efficient and secure distributed networks of cooperating sensors actuators and other smart devices and objects. This should enable person/object and object/object Internet-based communications opening a new range of Internet enabled services. The key challenges of the* **architecture** *are to move beyond the sector specific boundaries of the early realisations of the "Internet of Things," to cope with the heterogeneity of the underlying technologies, and to enable integration of the novel set of supported services with enterprise business processes.*

It is clearly seen that the focus is on architecture models while technological foundations is vague and may cover several different aspects: network technology, edge devices, software engineering.

On the other hand, other Challenges and Objectives may be related to IoT even if it is not explicitly mentioned, including (but not limited to):

Objective ICT-2011.1.2 Cloud Computing, Internet of Services and Advanced Software Engineering: *The objective focuses on technologies specific to the networked, distributed dimension of software and access to services and data. It will support long-term research on new principles, methods, tools and techniques enabling software developers in the EU to easily create interoperable services based on open standards, with sufficient flexibility and at a reasonable cost.*

Objective ICT-2011.1.4 Trustworthy ICT: *The objective is a trustworthy Information Society based on an ecosystem of digital communication, data processing and service provisioning infrastructures, with trustworthiness in its design, as well as respect for human and societal values and cultures. Projects must ensure strong interplay with legal, social and economic research in view of development of a techno-legal system that is usable, socially accepted and economically viable.*

Objective FI.ICT-2011.1.7 Technology foundation: Future Internet Core Platform: *The design, development and implementation of a generic, trusted and open **network and service Core Platform** making use of and integrating advanced Internet features supporting the uptake of innovative "smart applications." This includes the specification of open standardised interfaces from this Core Platform to use case-specific instantiations addressed by projects under Objective 1.8. A major aim is to offer Core Platform functionalities that can be generically reused in multiple usage contexts to support "smart applications" of various natures.*

Objective FI.ICT-2011.1.8 Use Case scenarios and early trials: *The work focuses on vertical use case scenarios whose intelligence, efficiency, sustainability and performance can be radically enhanced through a tighter integration with advanced Internet-based network and service capabilities.*

The target use cases should cover innovative applications scenarios with high social or economic impact making use of advanced Future Internet capabilities. Without being restrictive, examples of such target use cases include systems for utilities like the electricity grid, for traffic and mobility management, for health, and for ubiquitous access to networked digital media.

Objective ICT-2011.5.1: Personal Health Systems (PHS): (*a*) *Personal Health Systems for remote management of diseases, treatment and rehabilitation,* outside hospitals and care centres. Research will support innovations at system level and at component level if required. Solutions will be based on closed-loop approaches and will integrate components into wearable, portable or implantable devices coupled with appropriate platforms and services.

The EC has deviated the focus of network technology to other objectives rather than the IoT's one. This means that many useful results for IoT may be being produced without knowledge of the IoT community and maybe some technological gaps are not being looked upon.

While the IERC promotes the transversal coordination of IoT related projects **there is room for another transversal action crossing different objectives and challenges.**

9.4 Discovery and Search Engines

9.4.1 Real-time Search for Real-world Entities

The search for documents (Web pages, videos, blog entries, etc.) on the Web has become one of its most popular services. It is expected that the search for real-world entities will become equally important. The vast number of sensors that will be connected to the Web, the anticipated frequency of changes in sensor readings, and the requirement to search for the real-time state of real-world entities would all place huge demands on a real-world search engine. Römer et al. [18] survey prominent systems that support the discovery of real-world entities. While some of them have been explicitly designed with entity discovery in mind, others support the discovery of sensors with certain properties, but their underlying mechanisms would also apply to entity discovery. Yet another class of systems is concerned with real-time searches for dynamic documents in the current Web — a problem closely related to discovering entities based on their current state.

Dyser is a prototypical search engine for the Web of Things, developed by the authors [18]. It allows real-world entities (i.e., people, places, and objects) to be searched by their current state. As the expected audience of this search engine consists of average Web surfers rather than domain experts, it can be argued that users will most likely not be interested in searching for sensors with a specific reading, but for entities in the real world with a specific current state. So rather than searching for loudness sensors with a current reading below 30 dB, it is assumed that users will instead be interested in searching for places that are currently quiet. The state of an entity is determined by the current states monitored by its associated sensors.

The state a sensor outputs is inferred from its readings. In this model, there are two key elements: sensors and entities. Each sensor and each entity has a virtual counterpart, a Web resource, identified by a URL and accessible using HTTP. For all of these Web resources, there is always an HTML representation, which is called the *sensor page* and the *entity page*, respectively. In addition to unstructured text, they also contain structured information, for example, the type of sensor or its possible readings. The relationship between sensors and entities is many-to-many, i.e., one sensor can be associated with multiple entities, and one entity can be associated with multiple sensors. The proposed search language is quite simple and based on that of popular Web search engines. The user specifies a list of keywords and properties that have to be fulfilled by possible results. For example, when looking for quiet Italian restaurants, the user could search for *"italian restaurant loudness: quiet."* The first two elements of the query are static keywords, which have to appear on the entity page. The last element of the query specifies a (dynamic) property that is determined automatically by an associated loudness sensor. *Dyser* will return a ranked list of entities that currently match the search term. From there, the user may browse to the corresponding entity pages and in turn to the associated sensor pages, for further information.

Although a central index consisting of static metadata for sensors and entities has been built, *Dyser* does not rely on a central sink to which the readings of all the participating sensors are streamed in real time. This means there is no global view of the world and thus to answer a query, the search engine needs to contact relevant sensors (i.e., sensors that could possibly read the searched state) at the time a query is posed, in order to determine whether they currently match the query. Entities whose associated sensors do not read the searched state are excluded from the result set, which *Dyser* returns to the user as soon as enough matches have been found. Indexing current sensor readings is not an option, as the index would be outdated as soon as it was built due to the anticipated frequency of changes in sensor readings.

Even though this query resolution process can be optimized by contacting multiple sensors in parallel, it is not scalable with respect to network traffic — given that the number of possible results is significantly larger than the number of actual results, many sensors would be contacted unnecessarily by the search engine when processing a single query. However, instead of contacting the sensors in an arbitrary sequence, they could be contacted in an order that reflected the probability that they currently matched the query. This approach is called sensor ranking and it enables scalability, provided one can order the list of relevant sensors sufficiently well. For this, it is necessary to be able to estimate the probability of a sensor matching a query with sufficient accuracy. Prediction models have been used for this purpose, which return the probability of a sensor reading a specific value at a given point in time.

The classification of the systems supporting entity discovery according to their design space is shown in Table 9.5.

9.4.2 Discovery Service Architectures for the IoT

Today's use cases in which RFID is applied for identification and tracking of objects are mostly confined to manufacturing or companies that only implement RFID together with selected supply chain partners. In the future IoT, data of real world objects and events will be available globally and in vast amounts. These data will be stored in widely distributed, heterogeneous information systems, and will also be in high demand by business and end user applications. Therefore, a discovery mechanism that allows accessing such data is needed, even if its location and form of storage are unknown to the requester. It is conceived that so called Discovery Services (DS) will respond to such requests by returning a list of corresponding data providers.

Regarding the design of DS, there are several design decisions to be made. Evdokimov et al. [19] present five important approaches that propose distinct architectures for DS, and compare their characteristics. Qualitative attributes for DS architectures have not been comprehensively studied in research so far. The novelty of those architectures, and the fact that most of them are currently only available as architecture proposals or pilot implementations, makes a quantitative evaluation and benchmarking difficult, if not impossible. Therefore, a quality evaluation framework for conducting an evaluation of the existing DS architectures has been constructed.

They described requirements and analyzed the following five approaches for implementing DS in the IOT — EPCglobal, BRIDGE, Afilias DS,

Table 9.5. Classification of systems supporting entity discovery according to their design space dimensions.

Dimension	Snoogle	MAX	OCH	DIS	GSN	Sense Web	RT Web Search	Dyser
Query type	Ad hoc	Ad hoc	Continuous	Both	Ad hoc	Ad hoc	Ad hoc	Ad hoc
Query language	Key-words	Key-words	Key-words	Image	Key-words	Key-words + geo	Key-words	Key-words
Query scope	Local	Local	Global	Local	Global	Global	Global	Global
Query time	Real-time	Real-time	Real-time	Both	Real-time	Real-time	(near) real-time	Real-time
Query accuracy	Heuristic	Heuristic	Heuristic	Heuristic	Exact	Exact	Exact	Exact
Query content	Pseudo static	Pseudo static	Dynamic	Dynamic	Static	Static	Dynamic	Dynamic
Entity mobility	Mobile	Mobile	Mobile	Mobile	Mobile	Mobile	—	Mobile
Entity state	Categorial	Categorial	Categorial	Continuous	Categorial	Both	Text (categorial)	Categorial
Target users	End users	End users	End users	End users	Experts	Experts	End users	End users

ID@URI, and DHT-P2P. For comparing and evaluating these innovative architectures, they developed a quality framework based on the ISO/IEC 9126 standard and a literature review. This framework does not only provide a structured analysis tool, it can also be used by software developers, consulting companies, and service providers to individually evaluate different solutions for DS in the IoT, enabling a deeper understanding, improvement, or mutual integration of the approaches. Subsuming the current state of DS designs, it can be stated that the EPCglobal approach is still in development. Components of the relatively mature Afilias approach will probably be integrated into the EPCglobal standard, while the prototype of the BRIDGE project is very similar to the currently discernible EPCglobal approach. All three of them share the same advantages and disadvantages. Two very contrasting approaches are ID@URI and DHT-P2P. Both have several advantages and disadvantages: ID@URI is easy to deploy, but is dependent on the manufacturer who alone has the responsibility for providing object information in the supply chain. DHT-P2P delivers a very scalable and flexible solution. However, it remains unclear, who should bootstrap such an approach and how support for trackability and analyzability can be improved.

Limitations of the work are primarily the lack of real-world tests.

9.4.3 Comments

Two different approaches concerning real-time search for real-world entities and discovery services architectures for the IoT have been summarized. They show that a great effort is being put into this issue and much more is needed. It is important to realise that the performance of IoT applications and services is highly dependent on discovery services that may lay hidden in the overall architecture. The characteristics and/or requirements of discovery services need to be made explicit to plan a new IoT service development and deployment. This reinforces the need for **an all-inclusive harmonised IoT model and to develop a methodology for integrated roadmap building**.

9.5 Security and Privacy

9.5.1 Roles and Rights Management Concept

Brandherm et al. [20, 21] present a prototypical implementation (as a demonstrator) of an authorized access to a so-called Digital Product Memory (DPM) by means of a roles and rights management concept with identification by eID card. This approach has been demonstrated as an example on an instrumented medicament blister.

One possibility to identify authorized persons is given by the new electronic identity card (eIDcard). It has the following three main applications: (1) the ePass application (this application is only available for mandatory tasks), (2) mutual online identification and (3) qualified electronic signature. In this case the interest lays on the mutual online identification. It provides a so-called pseudonym identification, which enables service providers to recognize a known eIDcard without reading personal data. Besides the eIDcard the secret eIDPIN is necessary for online identification. Thereby the identification is not only tied to the eIDcard but also to the owner (proof of possession and knowledge; in comparison to the proof of knowledge and knowledge in the case of a login and a password). PACE (Password Authenticated Connection Establishment) is a cryptographic protocol which was developed to provide a secure knowledge-based authentication mechanism for contact-less chips. In Figure 9.12 the usage of the protocol PACE to establish a secure end-to-end connection between service and eIDcard is sketched: (1) The service owns a certificate issued by a public authority. It contains information about the identity of the service provider. (2) The user enters his eIDPIN and grants access to his eIDcard. (3) PACE. (4) The card verifiable certificate is sent to the chip of the eIDcard. (5) + (6) the service authenticates against the chip and the chip against the service.

The aforementioned service would be a roles and rights model server. Once successfully identified at this service the user (in this case the patient)

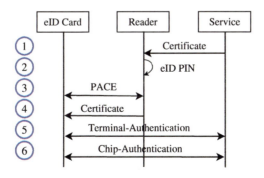

Fig. 9.12 Password authenticated connection establishment (PACE).

has access to the decryption key of the blister. The patient can grant other users (such as his physician or his pharmacist) access to this decryption key (without giving them the key) such that they have access to the DPM of the blister too.

But in the case of a pharmacy there is more than one pharmacist employed and the employed pharmacist may change over time. Here it would be useful to grant access to the role "pharmacist in this pharmacy."

Unfortunately the eIDcard doesn't manage user roles. Therefore a rights and roles model is needed. In this example scenario the employees of Pharmacy X have to be registered at a certified rights and roles server as in the role "pharmacist of X." Now the patient can grant the role "pharmacist of X" — and thus all matching people — access to the DPM.

9.5.2 Securing Cloud from DDOS (Distributed Denial of Service) Attacks

Cloud computing is one the most important technological pillars for IoT services deployment.

As cloud computing has taken hold, according to Bakshi et al. [22] there are six major benefits that have become clear: (i) Anywhere/anytime access — it promises "universal" access to high-powered computing and storage resources for anyone with a network access device. (ii) Specialization and customization of applications — it is a platform of enormous potential for building software to address a diversity of tasks and challenges. (iii) Collaboration among users — cloud represents an environment in which users can develop software-based services and from which they can deliver them. (iv) Processing power on demand — the cloud is an "always on" computing resource that enables users to tailor consumption to their specific needs. (v) Storage as a universal service — the cloud represents a remote but scalable storage resource for users anywhere and everywhere. (vi) Cost benefits — the cloud promises to deliver computing power and services at a lower cost.

Looking at the list of benefits, they actually highlight which are the top three concerns organizations have with cloud computing. It revolves around understanding how:

- Software As A Service (SaaS) provides a large amount of integrated features built directly into the offering with the least amount of

extensibility and a relatively high level of security. Since the user can only access or modify the data on the pre-defined application the underlying security issues are not of much concern.
- Platform As A Service (PaaS) generally offers less integrated features since it is designed to enable developers to build their own applications on top of the platform and is therefore more extensible than SaaS by nature, but due to this balance trades off on security features since user is responsible for program security and security issues.
- Infrastructure As A Service (IaaS) provides few, if any, application-like features, provides for enormous extensibility but generally less security capabilities and functionality beyond protecting the infrastructure itself since it expects operating systems, applications and content to be managed and secured by the consumer.

While protecting data from corruption, loss, unauthorized access, etc. are all still required characteristics of any IT infrastructure, cloud computing changes the game in a much more profound way. In addition, cloud computing goes hand-in-hand with the concept of virtualisation.

Virtualisation refers to the abstraction of logical resources away from their underlying physical resources in order to improve agility and flexibility, reduce costs and thus enhance business value. In a virtualised environment, computing environments can be dynamically created, expanded, shrunk or moved as demand varies. The server virtualisation is accomplished by the use of a hypervisor to logically assign and separate physical resources. The hypervisor allows a guest operating system, running on the virtual machine, to function as if it were solely in control of the hardware, unaware that other guests are sharing it. Each guest operating system is protected from the others and is thus unaffected by any instability or configuration issues of the others. Hypervisors are becoming a ubiquitous virtualisation layer on client and server systems. Virtualisation is extremely well suited to a dynamic cloud infrastructure, because it provides important advantages in sharing, manageability and isolation (that is, multiple users and applications can share physical resources without affecting one another). Virtualisation allows a set of underutilized physical servers to be consolidated into a smaller number of more fully utilized physical servers, contributing to significant cost savings.

Intrusion Detection System (IDS) is installed on the virtual switch which logs the network traffic in-bound and out-bound into the database for auditing. The packets are examined in real-time by the IDS for a particular type of attack based on predefined rules. The rules are defined based on well known attack strategies by the intruders. The IDS could determine the nature of attack and is capable of notifying virtual server the amount security risks involved. The virtual server on examining the security risks involved performs emergency response to the attack by identifying the source IP addresses involved in the attack could automatically generate the access lists that would drop all the packets received from that IP. If the attack type is *DDoS attack*, the botnet[2] formed by all the zombie machines are blocked. The virtual server then responds to the attack by transferring the targeted applications to virtual machines hosted in another datacenter. Router automation would immediately re-route operational network links to the new location. Hence, the firewall located at the new server will block all the IP addresses that attacker used and if any genuine user is trying to connect to the server, he will be redirected to the new server.

9.5.3 Comments on Security

The summarized articles show two different kinds of security concerns: (i) user authentication and data protection; (ii) attacks protection. The former has been addressed by specific protocol and cryptography and the latter by an intrusion detection system and migration of services. The articles provide just a glimpse of the IoT security universe. It is worth noting that, as has already been mentioned, much of the work concerning security are being developed by projects answering other objectives rather than that focused on IoT. The need of establishing a transversal effort through different challenges and objectives is reinforced.

9.5.4 Do People Care About Privacy?

It is widely accepted that people neither want nor like to disclose details of their private lives. Such statement needs to be taken with care.

---

[2] A botnet (also known as a zombie army) is a number of Internet computers that, although their owners are unaware of it, have been set up to forward transmissions (including spam or viruses) to other computers on the Internet. Any such computer is referred to as a zombie — in effect, a computer "robot" or "bot" that serves the wishes of some master spam or virus originator.

Michael et al. [23] conclude that item level tracking, for instance, comes with its own endowed advantages and benefits for some organizations within a retail supply chain context but may not be desirable for other application areas. A level of harmonisation needs to be reached between the level of required visibility in a given service and adhering to a consumer's right to informational privacy. Solutions can be devised and built-into the design of a service to overcome such challenges; they just need to be innovative. **If a consumer perceives that the value proposition to them of using a given technology outweighs any costs they may experience, then they are likely to adopt the technology.**

Michahelles et al. [24] arrive at a more radical conclusion reporting that Andrea Girardello (ETH Zurich) challenged the conventional wisdom about privacy when presenting TwiPhone, a mobile app that posts mobile phone event data, such as time and caller ID, as well as SMS communication, including text contents, to Twitter. According to Girardello, this essentially meaningless application, which seems nightmarish for privacy advocates, has surprisingly been downloaded by several thousand Android users. TwiPhone is also regularly used by several hundred people whose conversations can be publicly retrieved on Twitter using the #twiphone hashtag. **Do users not care about privacy anymore, or are they just unaware? Is privacy becoming an optional feature?**

9.5.5 Comments on Privacy

It is clearly seen that people behaviour is quite unpredictable and certainly diversified when privacy comes to the fore. There is also no doubt that the IoT community has to define how to deal with privacy issues that do not seem manageable by means of standards. On the one hand a strict privacy framework ruled by standards may inhibit the development and/or adoption of IoT technology: costs rise and privacy violations would be amenable to law suits. On the other hand there are privacy violations that cannot be left unpunished. A privacy legal framework needs to be established. The terms of such framework remain to be detailed but it seems to be a consensus about the opt-in concept. In her opening speech at the ICT2010 conference in Brussels, EP Vice-President MEP Silvana Koch-Mehrin pointed out that the IoT must be an "Internet of the People" whereby every person must be able to opt-in (not opt-out).[3] This can only happen if people are informed — which is currently not the case. Therefore, it is necessary to take a first step in establishing a laymen-language IoT Wiki in which important principles, concepts and technologies will be explained in simple language by experts coordinated by the IERCluster.

9.6 Application Areas and Industrial Deployment

The bibliographical review produced a large number of papers that were classified as "application areas and industrial deployment" but many of them fail to answer adequately the question: *"What is it that requires access to an Internet structure to fulfil the application or service requirements?"*

In many cases IoT is just a background scenario and the term is being employed just as a buzzword to make it easier to have the paper published. Among those real IoT applications it is possible to identify some characteristics that may help to understand what is going on.

There are applications in which the information associated to an object increases along the time as a historical record of the major events of its lifetime. It we take a pharmaceutical product it is possible to have information about manufacturing, transport, storage conditions, validity time, adverse drugs reactions, etc. Different kinds of information are of interest for different people with different roles: the pharmaceutical industry; the pharmacist; the physician; the patient. This characteristic is present not only in the healthcare area but also in industrial goods, agriculture and the food industry. In general the high societal value of such applications is easily recognisable. Good examples of this kind of application are the already mentioned works of Brandherm et al. [20, 21] for healthcare or the Stephan et al. [25] work concerning product-mediated communication.

There are applications designated as closed-loop applications in which an end user or an object triggers the process by accessing the information present in data carrier and the final action will be on the accessed object or the end user. Healthcare and ambient assisted living benefits significantly from this kind of applications as has been demonstrated, for example, by the works of Dohr et al. [26] and Jara et al. [27]. On the one hand the societal value is remarkable

---

[3] For the entire speech: http://mfile3.akamai.com/19611/wmv/ict2010.clients.telemak.com/vod/Morning_session_2709_Silv.asx.

but it is necessary to emphasize also the economical value arising from, for example, the decrease in the number of visits to hospitals and avoidable deaths.

Another distinguishing IoT characteristic is context awareness. Context awareness is a fundamental component for informational assistance and intelligent environment behaviour systems requiring constant intelligence rule updates. Important theoretical advances validated by means of testbeds have been reported by the research teams of Portmann [28], Kranz [29], Garrido [30] and Vlist [31].

There are applications for which is difficult to recognize importance or relevance. Any analysis would be influenced by taste, prejudice or cultural values. They target niche communities as youngsters, music or game lovers. However, such applications are good vehicles to drive technology to its limits and should not be discouraged for any reason. One the one hand it does not seem that IoT needs a killer application but the identification of applications that will bring more benefits for the society. On the other hand benchmarks and testbeds to identify scalability and interoperability issues are a must.

Looking at the whole universe of applications it possible to devise, broadly speaking, four categories: non-IoT, Intranet of Things, Web of Things and IoT. It is difficult but extremely important to be able to define precisely each one of them and identify similarities and dissimilarities. Similarities can lead to common developments and avoidance of wasting research effort. Dissimilarities may lead to optimization of performance and cost.

9.7 Governance and Socio-economic Ecosystems

9.7.1 Governance and Profit

Schneier [32] proposed the following taxonomy of social networking data at the Internet Governance Forum meeting in November 2009.

Service data is the data you give to a social networking site in order to use it. Such data might include your legal name, your age, and your credit-card number.

Disclosed data is what you post on your own pages: Blog entries, photographs, messages, comments, and so on.

Entrusted data is what you post on other people's pages. It's basically the same stuff as disclosed data, but the difference is that you don't have control over the data once you post it — another user does.

Incidental data is what other people post about you: a paragraph about you that someone else writes, a picture of you that someone else takes and posts. Again, it's basically the same stuff as disclosed data, but the difference is that you don't have control over it, and you didn't create it in the first place.

Behavioral data is data the site collects about your habits by recording what you do and who you do it with. It might include games you play, topics you write about, news articles you access (and what that says about your political leanings), and so on.

Derived data is data about you that is derived from all the other data. For example, if 80 percent of your friends self-identify as gay, you're likely gay yourself.

Such taxonomy has not been proposed having IoT in mind. However, its concepts may be useful to be adapted or to be taken into account at least in scenarios like the one proposed by Guinard et al. [33] in which they propose a platform that enables people to share their Web-enabled devices so that others can use them. They illustrate how to rely on existing social networks and their open APIs (e.g., OpenSocial) to enable owners to leverage the social structures in place for sharing smart things with others.

In addition to the IoT governance aspects, business models are also of relevance. As already mentioned, cloud computing is one of the most important IoT technological foundations. Goiri et al. [34] report that cloud federation has been proposed as a new paradigm that allows providers to avoid the limitation of owning only a restricted amount of resources, which forces them to reject new customers when they have not enough local resources to fulfil their customers' requirements. Federation allows a provider to dynamically outsource resources to other providers in response to demand variations. It also allows a provider that has underused resources to rent part of them to other providers. Both things could make the provider to get more profit when used adequately. This requires that the provider has a clear understanding of the potential of each federation decision, in order to choose the most convenient depending on the environment conditions.

9.7.2 Comments

It is remarkable to verify that the bibliographical review produced a very small and disappointing number of papers on governance that is one of the

EC priorities. It could be argued that there is a conflict between the EC's and the IoT community's view. However, it is important to realise that this bibliographical review was restricted to IEEE papers. The IEEE community consists mainly of technology people trying to solve technological problems. The governance concerns are being dealt with in different fora. It is then necessary to develop another transversal action through different communities to promote convergence and avoid conflicts between governance requirements, technology possibilities and business targets.

These two works put in evidence once again the need to create an all-inclusive model that take into account IoT dimensions other than the technological: governance, business models, security and privacy, human factors. The model needs to be evolvable to incorporate new requirements and characteristics not foreseeable today. The model has to lead to a methodology for roadmapping.

9.8 Contribution to an Iot Global Vision: Opportunities and Challenges

The seven prior sections are completely agnostic in terms of geography, culture or ethnicity. In this section it will be presented an IoT view considering the reality of South America.

South America has an area of 17,840,000 square kilometres (6,890,000 sq mi), or almost 3.5% of the Earth's surface. As of 2008, its population was estimated at more than 390,000,000. South America ranks fourth in area (after Asia, Africa, and North America) and fifth in population (after Asia, Africa, Europe, and North America). Table 9.6 shows the top 10 Internet countries in Latin America where it can be seen that 7 are from South America. It can also be seen that there is an enormous diversity in terms of absolute numbers of Internet penetration, but almost all countries present a very expressive growth rate.

A key point to be taken into account is that **there is no Strategic Research Agenda in South America**. The expressive growth that can be observed does not fulfil challenges and objectives of any work programme. Sometimes there are actions that respond to a need of some very specific stakeholder. For example, in Brazil, there is a program called Brasil-ID whose objective is to deploy a RFID infrastructure along the main roads to monitor goods transportation

Table 9.6. Latin American Internet usage — top 10 Internet countries [35].

Latin America countries	Population (Est. 2010)	Internet users, Latest data	% population (penetration)	User growth (2000–2010)	% users in Table
Argentina	41,343,201	26,614,813	64.4%	964.6%	13.3%
Brazil	201,103,330	75,943,600	37.8%	1,418.9%	37.9%
Chile	16,746,491	8,369,036	50.0%	376.2%	4.2%
Colombia	44,205,293	21,529,415	48.7%	2,352.1%	10.8%
Costa Rica	4,516,220	2,000,000	44.3%	700.0%	1.0%
Dominican Republic	9,823,821	3,000,000	30.5%	5,354.5%	1.5%
Ecuador	14,790,608	2,359,710	16.0%	1,211.0%	1.2%
Mexico	112,468,855	30,600,000	27.2%	1,028.2%	15.3%
Peru	29,907,003	8,084,900	27.0%	223.4%	4.0%
Venezuela	27,223,228	9,306,916	34.2%	879.7%	4.7%

across the states borders in the country. The stakeholders are the State Financial Secretaries and the Ministry of Science and Technology. It is an important project but it is not part of any national or regional ICT work programme.

South America needs a roadmap for its growth. Not only in IoT but for all technological fields.

It must be underlined that, before valid and comparable roadmaps can be identified, it is mandatory to create a methodology that harmonises the development of integrated roadmaps taking into account all potential stakeholders and the relationships between them considering all the necessary aspects and dimensions involved. This "reference modelling" can be best explained in using an **example** of *traffic management*.

Let's assume we want to produce a roadmap for a city traffic management service that should receive multimedia traffic information from mobile devices like cameras and web cameras in cars and motorcycles, send this information to processing centres that would return complete updated real time traffic maps to guide the driver in choosing good alternative paths in the streets, roads, etc. This is an IoT service that does not use RFID tags but other input devices. It could also use sensors deployed along the streets to perceive the traffic in real time.

Figure 9.13 shows the described scenario mapped into the IoT Technological Plane. This path represents the kinds of technologies, applications, users, etc. that need to be involved to build the service.

If we build other planes, like, for example, an Integrated Management and Control Plane, it will be possible to identify the technologies used or necessary to do the job, and how to integrate them.

9.8 Contribution to an IoT Global Vision: Opportunities and Challenges

Components								
1	2 / 3	4 5		6 7		8	9 / 10	11
Environment and contexts	Human users/ Devices	Client apps.	Mobile access networks	Back bones	Fixed access networks	Server applications	Human users/ Devices	Environment and contexts
Streets of a city Heavy traffic Building	Drivers Motor cycles Webcams Small monitors	Client apps. atcars, motor cycles	WiMax 3GLTE	Internet access providers Backbone providers MPLS, GMPLS, DWDM Traffic eng.	IP MPLS Metro Ether net SDH	Databank Maps Visualization Graphs	Service staff Global admin. Integrated maps	City and roadservice Traffic admin. Emergency admin. Catastrophic scenarios

Data capture ⟶ Actuation

Fig. 9.13 Example of a technological IoT scenario.

According to Figure 9.13, the scenario has 11 components on the IoT technological plane. It is important to realise that some of these components may represent technological alternatives, while others may be technological requirements. For example, for component 5 (WiMax, 3G, LTE) each one of the listed technologies may be considered as a solution to deploy the service or it may say that the service requires the possibility to be accessed by all the listed technologies. These two views may lead to different roadmaps.

Each of these components is influenced by other dimensions: Security, Business Processes, Integrated Management and Control, Regulations, and Human Factors.

Continuing to use component 5 as an example, Table 9.7 lists some hypothetical characteristics for two cities in South America: São Paulo (Brazil) and Medellin (Colombia).

Clearly, the table presents characteristics from planes other than the technological plane:

- Equipment availability: technology plane
- Geographical coverage: business process plane
- Cost: business process and human factors planes
- National regulation: regulations plane

The roadmaps for the two cities are quite different. In São Paulo, for any technology the geographical coverage is quite low. It is necessary to provide

Table 9.7. Hypothetical characteristics for two cities in South America: São Paulo (Brazil) and Medellin (Colombia).

Technology	Equipment availability	Geographical coverage	Cost	National regulation
São Paulo, Brazil				
WiMAX	Yes	30%	High	Yes
3G	Yes	40%	Low	Yes
LTE	No	N.A.	N.A.	N.A.
Medellin, Colombia				
Technology	Equipment availability	Geographical coverage	Cost	National regulation
WiMAX	Yes	20%	High	No
3G	Yes	90%	Medium	Yes
LTE	No	N.A.	N.A.	N.A.

incentives for the operators to invest in infrastructure. There is also a problem concerning the high cost of WiMax equipment and the reason may associated to high taxes. In Medellin, the adoption of the 3G technology could allow an almost immediate deployment of the service with high geographical coverage. The lack of national regulation for the WiMax technology may prevent to offer the service to distant populations and it is necessary to find ways to decrease the cost of such equipments.

It is obvious that such simplistic and ad-hoc analysis cannot be employed systematically to produce realistic and useful roadmaps, but it indicates the need to adopt a holistic approach and consider the specificities of different realities.

IoT is much more than technology and it does not exist without business models. It is necessary to perform a critical review of proposed IoT businesses models. The universe of any bibliographical review should encompass the universe of socio-economic ecosystems to bring the needs, expectations and opportunities related to different realities to be taken into account. In particular, when we consider South America, i.e., lower-middle income countries, a special attention has to be given to the "Bottom of the Pyramid." The so-called "Bottom of the Pyramid" is a global market, usually despised by the companies, where 4 billion people earn up to US$4 a day. In order to enable the appropriation of new technologies, as IoT, by such market innovative business models have to be proposed. Associating the information obtained from the bibliographical review with the inputs from an intense international dialogue socio-economic scenarios focused on different geographical realities can be devised.

References

[1] New Network Architectures: The Path to the Future Internet, Tania Tronco (Ed.), *Studies in Computational Intelligence 297*, Springer Verlag, 2010, ISSN 1860-949X.
[2] The Internet of Things, ITU Internet reports 2005.
[3] EPCglobal architecture model available in http://www.gs1.org/gsmp/kc/epcglobal.
[4] CASAGRAS Final Report available in http://www.rfidglobal.eu/.
[5] N. Koshizuka and K. Sakamura, Ubiquitous id: Standards for ubiquitous computing and the Internet of Things, *Pervasive Computing, IEEE*, 9(4), pp.98–101, 2010.
[6] O. Akribopoulos, I. Chatzigiannakis, C. Koninis and E. Theodoridis, A Web Services-oriented architecture for integrating small programmable objects in the Web of Things. In *Developments in E-systems Engineering (DESE)* 2010, pp. 70–75, 2010.
[7] L. Tan and N. Wang, Future Internet: The Internet of Things, In *Advanced Computer Theory and Engineering (ICACTE), 2010 3rd International Conference on*, 5, pp. V5-376–V5-380, 2010.
[8] G. Kortuem, F. Kawsar, D. Fitton, and V. Sundramoorthy, Smart objects as building blocks for the Internet of Things, *Internet Computing, IEEE*, 14(1), pp. 44–51, 2010.
[9] D. Guinard, Towards opportunistic applications in a Web of Things, In *Pervasive Computing and Communications Workshops (PERCOM Workshops), 2010 8th IEEE International Conference on*, 2010.
[10] D. Guinard, V. Trifa, S. Karnouskos, P. Spiess and D. Savio, Interacting with the SOA-based Internet of Things: Discovery, query, selection, and on-demand provisioning of Web Services. *Services Computing, IEEE Transactions on*, 3(3), pp. 223–235, 2010.
[11] G. Nain, F. Fouquet, B. Morin, O. Barais and J.-M. Jézéquel, Integrating IoT and IoS with a component-based approach. In *Software Engineering and Advanced Applications (SEAA), 2010 36th EUROMICRO Conference on*, pp. 191–198, 2010.
[12] Information about the IoT-A project is available in http://www.iot-a.eu/public.
[13] N. Glombitza, D. Pfisterer, and S. Fischer, LTP: An efficient Web Service transport protocol for resource constrained devices, In *Sensor Mesh and Ad Hoc Communications and Networks (SECON), 2010 7th Annual IEEE Communications Society Conference on*, pp. 1–9, 2010.
[14] F. Wang, Z. Wang, Y. Li and L. Zeng, Reliable multi-path routing with bandwidth and delay constraints, In *Multimedia Technology (ICMT), 2010 International Conference on*, pp. 1–6, 2010.
[15] J. Liu and W. Tong, Adaptive service framework based on Grey decision-making in the Internet of Things, In *Wireless Communications Networking and Mobile Computing (WiCOM), 2010 6th International Conference on*, pp. 1–4, 2010.
[16] E. Le Rouzic, Network evolution and the impact in core networks, In *Optical Communication (ECOC), 2010 36th European Conference and Exhibition on*, pp. 1–8, 2010.
[17] FP7 Work Programme 2001 — Cooperation — ICT, Available in http://cordis.europa.eu/fp7/wp-2011_en.html.
[18] K. Römer, B. Ostermaier, F. Mattern, M. Fahrmair and W. Kellerer, Real-time search for real-world entities: A survey, *Proceedings of the IEEE*, 98(11), pp. 1887–1902, 2010.
[19] S. Evdokimov, B. Fabian, S. Kunz and N. Schoenemann, Comparison of discovery service architectures for the Internet of Things. In *Sensor Networks, Ubiquitous, and Trustworthy Computing (SUTC), 2010 IEEE International Conference on*, pp. 237–244, 2010.

[20] B. Brandherm, J. Haupert, A. Kroner, M. Schmitz and F. Lehmann, Roles and rights management concept with identification by electronic identity card, In *Pervasive Computing and Communications Workshops (PERCOM Workshops), 2010 8th IEEE International Conference on*, p. 29, 2010.

[21] B. Brandherm, J. Haupert, A. Kroner, M. Schmitz and F. Lehmann, Demo: Authorized access on and interaction with digital product memories, In *Pervasive Computing and Communications Workshops (PERCOM Workshops), 2010 8th IEEE International Conference on*, p. 29, 2010.

[22] A. Bakshi and B. Yogesh, Securing cloud from DDOS attacks using intrusion detection system in virtual machine, In *Communication Software and Networks, 2010. ICCSN '10. Second International Conference on*, pp. 260–264, 2010.

[23] K. Michael, G. Roussos, G. Q. Huang, A. Chattopadhyay, R. Gadh, B. S. Prabhu and P. Chu, Planetary-scale RFID services in an age of uberveillance, *Proceedings of the IEEE*, 98(9), pp. 1663–1671, 2010.

[24] F. Michahelles, S. Karpischek and A. Schmidt, What can the Internet of Things do for the citizen? *Workshop at Pervasive Computing, IEEE*, 9(4), pp. 102–104, 2010.

[25] P. Stephan, G. Meixner, H. Koessling, F. Floerchinger and L. Ollinger, Product-mediated communication through digital object memories in heterogeneous value chains, In *Pervasive Computing and Communications (PerCom), 2010 IEEE International Conference on*, 2010.

[26] A. Dohr, R. Modre-Opsrian, M. Drobics, D. Hayn and G. Schreier, The Internet of Things for ambient assisted living, In *Information Technology: New Generations (ITNG), 2010 Seventh International Conference on*, pp. 804–809, 2010.

[27] A. J. Jara, F. J. Belchi, A. F. Alcolea, J. Santa, M. A. Zamora-Izquierdo and A. F. Gomez-Skarmeta, A pharmaceutical intelligent information system to detect allergies and adverse drugs reactions based on Internet of Things, In *Pervasive Computing and Communications Workshops (PER- COM Workshops), 2010 8th IEEE International Conference on*, p. 29, 2010.

[28] E. Portmann, A. Andrushevich, R. Kistler and A. Klapproth, Prometheus — fuzzy information retrieval for semantic homes and environments, In *Human System Interactions (HSI), 2010 3rd Conference on*, pp. 757–762, May 2010.

[29] M. Kranz, P. Holleis and A. Schmidt, Embedded interaction: Interacting with the Internet of Things, *Internet Computing, IEEE*, 14(2), pp. 46–53, 2010.

[30] P. C. Garrido, G. M. Miraz, I. L. Ruiz and M. A. Gómez-Nieto, A model for the development of NFC context-awareness applications on Internet of Things.

[31] B. J. J. van der Vlist, G. Niezen, J. Hu and L. M. G. Feijs, Semantic connections: Exploring and manipulating connections in smart spaces, In *Computers and Communications (ISCC), 2010 IEEE Symposium on*, pp. 1–4, 2010.

[32] B. Schneier, A taxonomy of social networking data, *Security Privacy, IEEE*, 8(4), pp. 88–88, 2010.

[33] D. Guinard, M. Fischer and V. Trifa, Sharing using social networks in a composable Web of Things, In *Pervasive Computing and Communications Workshops (PERCOM Workshops), 2010 8th IEEE International Conference on*, 2010.

[34] I. Goiri, J. Guitart and J. Torres, Characterizing cloud federation for enhancing providers' profit, In *Cloud Computing (CLOUD), 2010 IEEE 3rd International Conference on*, pp. 123–130, 2010.

[35] Source: Internet World Stats — www.internetworldstats.com

10

Virtualization of Network Resources and Physical Devices in Internet of Things Applications

Dr. Sangjin Jeong, Dr. Myung-Ki Shin and Dr. Hyoung-Jun Kim

Electronics and Telecommunications Research Institute (ETRI), Korea

10.1 Introduction

After developing the concept of Internet several decades ago, the Internet has become one of the crucial things in the daily life. However, because of the rapid changes of communication environment and emerging of new services, increasing the users' requirements for service level, and so on, the discussion about how to evolve the current Internet architecture is undergoing in the network communities. Future Internet is expected to provide revolutionary services, capabilities, and facilities that are difficult to support using existing network technologies [1]. Also, it is expected that Future Internet will overcome the limitations of the current networks and will facilitate various emerging services and applications, such as Internet of Things (IoT) applications by achieving several design goals and high-level capabilities that are necessary for realizing the Future Internet. Some of the investigated design goals require separate network environment, coexistence of multiple networks, abstraction of network resources, or aggregation and federation of multiple networks and resources. Also, promising technologies for realizing the design goals may be designed not bound to the current networking technologies, because the promising technologies will be developed in order to overcome the limitations of the current networks [2]. Network virtualization allows networks without interfering with the operation of other virtual networks while sharing the network resources among virtual networks. Since multiple isolated virtual networks can simultaneously coexist over Future Internet infrastructures, different virtual networks can deploy heterogeneous network technologies which may cause interference among virtual networks and it is possible to improve utilization of physical resources. Also, the abstraction property enables to provide standard interfaces for accessing and managing the virtual network and resources and helps on supporting seamless migration and update the capability of virtual networks. Finally, network virtualization can aggregate multiple virtual resources into a single virtual resource in order to obtain increased capabilities. Therefore, network virtualization is a key technology that can enable the partitioning, isolation, abstraction, and aggregation of networks and can be used to realizing Future Internet.

10.2 Overview of Network Virtualization

The main objectives of virtualization are to create multiple logical instances of the resources that can coexist, to separate the uses of the logical instances, and to simplify the use of the underlying resources by abstracting the characteristics and interacting with the resources with limited abstracted knowledge. The virtualization technology has been extensively studied for decades from desktop virtualization, application virtualization, system virtualization, link virtualization, storage virtualization to network virtualization. All the virtualization technologies above have relationship with network virtualization technology, but from the simplified view the network elements may be seen as systems with links. Thus, network virtualization would be expected to be realized on the basis of traditional virtualization technologies, especially system and link virtualization. The system virtualization is the ability to run an entire virtual system with its own guest OS over another OS or over a bare-machine. So, it allows multiple virtual systems with heterogeneous guest OSes to run in isolation on the same physical system. Each virtual system has its own set of virtual hardware and can be accessed independently. Thus, each virtual system can participate in constructing and providing independent networks. The virtual system can utilize consistent, normalized set of hardware regardless of the characteristics of physical hardware specification. Network virtualization allows dynamic creation and management of virtual networks over network infrastructures. These virtual networks may be heterogeneous and multiple separate virtual networks can be simultaneously coexisted over the network infrastructures.

The virtual networks can be served as non-virtualized networks without operational interference with other virtual networks while sharing the components of networks. Thus, multiple virtual networks can concurrently use a single physical network for multiple virtual networks and different virtual networks may use heterogeneous network technologies in the isolated and separate environment. Also, standardized set of interfaces between virtual networks can make it easier to provide virtual networks and improve portability. The provision of standardized interfaces can support seamless migration and update of the capability of virtual networks. Finally, utilization of physical resources can be increased by accommodating multiple virtual networks in a single physical resource.

A virtual network is a network of virtual resources where the resources can be separated from other virtual resources and their capabilities can be dynamically reconfigured. In other words, a virtual network is a logical partition of physical or logical networks and its capability is the same as or subset of the networks. Also, the virtual network may expand its capability by aggregating the capabilities of multiple networks. From the user's point of view, the virtual network can be seen as a non-virtualized network. A virtual resource is an abstraction of physical or logical resource and its partition and has the same mechanisms as the physical or logical resource. It can also inherit all existing mechanisms and tools for the physical or logical resource. In addition to the mechanisms above, a virtual resource has several interfaces to access and manage the virtual resource. These interfaces typically include virtual data plane interfaces, virtual configuration interfaces, and management interfaces [3, 4].

Virtual networks are completely isolated each other, so different virtual network may use different protocols and packet formats. When combined with programmability in network elements, users of virtual networks can program the network elements on any layers from physical layer to application layer. They can even define new layering architecture without interfering with the operation of other virtual networks. In other words, each virtual network can provide the corresponding user group with full network services similar to those provided by a traditional non-virtualized network. The users of virtual networks may not be limited to the users of services or applications, but may include service providers. For example, a service provider can lease a virtual network and can provide emerging services or technologies such as cloud computing service, and so on. The service providers can realize the emerging services as if they own dedicated physical network infrastructures. In order to facilitate the deployment of network virtualization, it may be necessary to provide control procedures such as creating virtual networks, monitoring the status of virtual network, measuring the performance, and so on.

Also, network virtualization can reduce the total cost by sharing network resources. One of motivation of network virtualization is to achieve better utilization of infrastructures in terms of reusing a single physical or logical resource for multiple other network instances, or to aggregate multiples of these resources to obtain more functionality. These resources can be not only network components, such as routers, switches, hosts, virtual machines, but also service elements, such as, operation and measurement services, instrumental services, and so on.

Network virtualization on its own is quite a useful technology, but to gain the greatest benefit it will be better to have a managed virtualization environment. Such management allows better control, monitoring, and analysis of the virtualized environments.

Network virtualization is required to be capable of providing multiple virtual networks those are isolated each other. Each virtual network may be created over the single physical infrastructure.

Figure 10.1 represents the logical relationship between virtual networks and physical infrastructures. Each virtual network can be simultaneously created over shared physical resources. Each virtual network is isolated each other and may be programmable to satisfy the user's demand on the functionality and amount. Users' demand is conveyed to virtual network management entity which is required to coordinate infrastructures so that appropriate virtual network is provided to the user. Virtual networks are generally managed by management entity to handle user's demand with real-time or scheduled.

10.2.1 Benefits and Overheads of Network Virtualization

Network virtualization may increase availability via maximizing utility of physical network infrastructures. However, availability becomes an issue because if one physical network resource, which can be part of several virtual

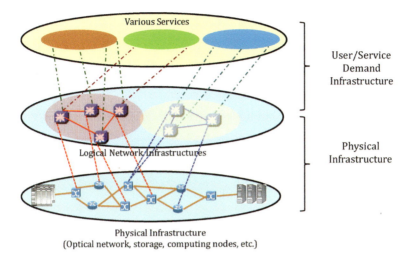

Fig. 10.1 Logical relationship of virtual networks and physical resources.

networks, were to go down, these virtual networks will be also influenced by the failed physical network resource. Another case is that where a virtual network management system fails or operates poorly. In this case all of the relevant virtual networks will also fail. Therefore most service providers for virtual network should implement failover mechanisms, such as failure-resistant configuration using cluster physical network elements, to prevent such outage. This remains necessary despite the fact that, each virtual network is guaranteed to isolate its network services and has management systems to handle physical network infrastructure.

Scalability is difficult to deal with under full utilization for capable physical network resources. A virtual network provides a way for corresponding user groups to support full network services, so that a service provider can lease a virtual network to provide emerging service to fit user demand using restricted physical resources. The problem is that while a service provider may implement a small scale virtual network, it may be that a fast growing network service could quickly require more resources. Users or service provider of the virtual network want to add as many physical resources as they need, but it may be difficult financially to lease physical network elements, such as routers, switches to meet this demand. So scalability and utilization are intertwined.

Additionally, virtual networks impose performance limitations by running an additional layer composed of programmable network elements above physical network infrastructures. The performance degradation depends on virtualization mechanisms for isolation and management, because many software technologies are involved to achieve better isolation and management system.

10.3 Challengens for Network Virtualization

This section investigates typical research problems for network virtualization [3].

10.3.1 Isolation

Legacy networks enable users to partially realize multiple network services over shared physical infrastructure in a way, which allows services to be affected by other coexisting network services. How to provide secure isolation among the network services is an important issue. The isolation has various aspects including security isolation, performance isolation, management isolation, and so on. For example, since multiple network services coexist over shared physical infrastructures, security problems in a service may be spread out over the whole networks and may cause performance degradation of other services. Network virtualization can provide complete isolation of any virtual network from all others, minimize the impact of behaviour of virtual networks to other networks, and support diversity of application, service, and architectures.

10.3.2 Performance

In legacy networks, the network service providers hardly offer resources encompassing the physical capability of the resources. However, by leveraging network virtualization, it is possible to provide high performance resources for users by logically aggregating multiple resources into single resource. Therefore, a logical network consisting of requested resources can guarantee users' performance requirements.

10.3.3 Scalability

In legacy networks, the scale of a network resource is restricted by the number of physical resources, so scaling out, i.e., adding additional physical resources to the network resource is one of simple methods to increase scalability. This approach can increase the management complexity of the network resource due to the increased number of physical resources. Network virtualization allows adding or aggregating additional logical resources to a virtual resource in order to provide increased throughput with less cost than adding physical resources.

10.3.4 Flexibility

Flexibility refers to the capability of building a system, and expanding it as need in order to adapt internal or external changes [6]. In legacy networks, network resources are composed of concrete network functions and equipments, so it is not easy to install new capabilities into existing architectures, especially without disruption of existing services. However, network virtualization can provide the capability of accepting different network architectures by aggregating logical resources into a virtual resource, which can increase flexibility of network resources.

10.3.5 Evolvability

If network providers want to deploy a new network technology, e.g., new network layer protocols in the current network architecture, they need to construct a separate testbed so that the behaviour of the new protocols does not affect the current services. After evaluating the new technology, the network provider deploys the new technology to their network. However, this approach may lose legacy support or backward compatibility and users may be reluctant to adapt the new feature because the new technology may not support their existing services. By utilizing network virtualization, the network providers can integrate legacy support by allocating the existing networks to a virtual network. The virtual network will guarantee that the existing services and technologies can remain unchanged.

10.3.6 Management

Since each virtual network is independent from other virtual networks, it has to be managed independently from other virtual networks. At the same time, the management system for the virtual network has to collaborate with the management system of physical infrastructure. Therefore, it is necessary to carefully define which part of management can be done by the management system of the virtual network, and how to align it with that of physical infrastructure. Moreover, if the isolation is not perfect, alignment with the management systems of other virtual networks also becomes necessary. When network providers want to examine the status of a virtual network such as topology of virtual network, allocation of physical resources to the virtual network, they may need to access multiple management systems. However, network virtualization can realize an integrated management system because virtualization layer or hypervisor can access information about both physical and logical resources.

10.3.7 Security

In the legacy network, failure or malfunction in one service or network can be spread over the whole network. Also, if a network element is compromised by a malicious user, the whole traffic traversing the element may be a target for attack. Since network virtualization ensures complete isolation among virtual networks, the security problems are not be spread over the whole networks. Also, secure isolation among virtual networks can prevent unauthorized access to a virtual network including its services and data.

10.4 Requirements of Network Virtualization

This section discusses the requirements of network virtualization.

10.4.1 Programmability

Network virtualization allows the creation of virtual networks by using not only local resources but also remote resources. A virtual network may be equipped with a programmable control plane so that users can use arbitrary network topologies, forwarding or routing functions, and customized protocols. In order to provide flexibility to the virtual network, it is required to implement new control schemes on virtual resources. Programmability can support flexibility in the control plane and make it possible to easily adopt new control schemes on virtual networks.

10.4.2 Topology Awareness

Since network virtualization allows aggregation of network resources distributed in network infrastructures, it is necessary to discover those and to connect the network resources. Further, when constructing or re-constructing a virtual network, optimization may be required for effective use of network infrastructure. For example, for the users who want low end-to-end delay, topology awareness may help to provide low delay by using the shortest route, but still provide a high bandwidth route for users who want high bandwidth only. Therefore, topology awareness is necessary so that network resources can seamlessly interact with each other during the construction of virtual networks.

10.4.3 Quick Reconfigurability

After creating virtual networks based on users' requirements, the capabilities of virtual networks need to be modified due to various reasons, for example the changes of users' requirements, the status of networks, policies of resource owners, and so on. Hence, each virtual network should be able to adjust its capability according to the changes of requirements and the reconfiguration should be quickly done in order to minimize service disruption. Thus, the network virtualization should offer a method that virtual network manage can easily and rapidly create virtual networks and dynamically reconfigure them.

10.4.4 Isolation

Virtual networks can be multiplexed over shared network infrastructures. However, this can be liable to restrict network performance and cause instability due to interference by other virtual networks. The network virtualization should be capable of providing the complete isolation among virtual networks, for example, performance isolation, security isolation, management isolation, etc.

10.4.5 Network Abstraction

A virtual network may accommodate a new architecture, so network resources should be controlled as abstract resources. Network abstraction allows hiding the underlying characteristics of network resources from the way in which other network resources, applications, or users interact with those network resources and separates infrastructure instances and control frameworks of network virtualization. Network abstraction also allows the controlled opening of networks functions. In other words, it allows selective exposure of key network functionalities in network infrastructures. Network abstraction will also allow infrastructure technology replacement with less impact to the framework of network virtualization. Therefore, it is necessary to support network abstraction methods to define levels of exposure and interfaces to other network resources.

10.4.6 Performance

Network virtualization is typically implemented by introducing virtualization layer or adaptation layer and the virtualization layer creates and manages virtual networks. Thus, the performance of the virtual networks is not as good as the non-virtualized network. Therefore, network virtualization should reduce the performance degradation due to the virtualization layer. Also, it is possible that a malfunctioned virtual network consumes most of physical network resources, which reduces the performance of other virtual networks due to network resource exhaustion. So, the network provider should have the capability of regulating the upper limit of bandwidth usage by each virtual network to maintain the overall throughput and performance.

10.4.7 Security

Since virtual networks created by network virtualization are completely isolated and independently managed, conventional security issues for non-virtualized networks should be also considered when implementing network virtualization in network resources. In addition to that, since network virtualization introduces additional layer such as virtualization layer or hypervisor and all virtual networks are managed by virtual network manager utilizing virtualization layer, a compromised virtual network manager may allow a malicious attacker to all virtual networks in physical network infrastructure. Therefore, how to secure management systems of virtual networks should be carefully considered before implementing network virtualization.

10.4.8 Management

To support diversified network services, the virtual networks should retain the capability of customizing network control and operations independent from those in the physical network or other virtual networks. At the same time, the virtual network wishes to avoid complex physical network operations that are fully dependent on the types of network layers and equipment vendors. To disengage the virtual network from the complexity of the physical network, the network virtualization should conceal a part of the physical network information and provides the simple interface for resource control to the virtual networks.

Each virtual network can be a flexible aggregation of physical network resources and logical network resources with arbitrary network topology. From this perspective, a number of associations, which are not only physical-to-physical resource but also physical-to-logical resource and vice versa have to be managed and are able to provide the visibility needed to understand all interconnections between physical and logical resources over the shared infrastructures. Thus, considering the rapidly changing of virtualized network environments, the visibility is essential for network management operations such as monitoring, fault detection, topology awareness, re-configurability, resource discovery/allocation/scheduling and customized control. Therefore, the aggregation of multiple resources requires change to be detected as it occurs.

10.4.9 Mobility

When the users' demands change or network performance falls below quality of service thresholds, the network resources such as storages, computing nodes, network switches, applications, etc. can be moved to other physical or virtual resources so that users' quality of service requirements are satisfied and stable network performance is maintained. Therefore, network virtualization should support two types of mobility. One is the conventional mobility managing the location change of mobile nodes in physical or virtual networks. The other one is mobility of the network resources including users among virtual networks. The network resources are composed of resources, system images and applications, etc. which are added to improve network performance or removed for load balancing or energy saving purpose. Users can be dynamically attached or reattached to one of the virtual networks depending on the application characteristics without any interruption of communication.

10.4.10 Wireless Virtualization

Since network resources are shared and limited in wireless networks, the exact analysis of throughput, delay, wireless link status, and so on is required to guarantee the performance in wireless virtualization, and efficient resource sharing mechanisms should be also developed. Meanwhile, multiple access points with different wireless networking technologies/protocols should operate together via the wireless virtualization. These multiple access points should appear to the users as a single access point. Therefore, the users are transparently connected to wireless networks while moving without knowing the change of the access points.

10.5 Applicability of Network Virtualisation

This section describes the applicability and use cases of network virtualization. The applicability of network virtualization may be classified into three broad categories according to the characteristics of virtualization: Namely isolation, aggregation, and federation.

From the viewpoint of isolation, network virtualization enables the complete isolation between each logical network partition. A typical use case for this category is that it is possible to create new business model by separating the conventional Internet service provider's role into network provider and service provider. The network provider creates customized virtual networks according to the service providers' requirements, such as network bandwidth, the number and functional capability of network elements in the virtual networks, total cost, and so on. The service provider can provide various services and applications for users without the burden of building its own network infrastructure. Moreover, the service provider can safely test an innovative pilot service or application that requires special features of network without affecting other existing services. From the perspective of network providers, they can increase the hardware utilization ratio and can reduce the operational cost of network infrastructure. From the viewpoint of service providers, it is possible to reduce the cost for building network infrastructure. Also, they can utilize the flexibility for the creation of network and application services.

The second applicability is aggregation. In the computing field, it is common to logically aggregate multiple computing resources into a single resource in order to support applications or services that require very high performance exceeding single resource's capability. Building a high performance computing node by clustering many smaller nodes would be the typical use case of resource aggregation. Similar to this use case, network virtualization can allow building a logical network element whose capability is hard to be supported by a single one. The logical network element can support various functions and can easily expand its capability by aggregating multiple network elements.

The third applicability is federation. By applying virtual network management to federation, a virtual network can comprise various types of resources from independently administered networks. Due to the heterogeneous nature of networks, a new federation management function is required to interact with different types of networks and to control the life cycle of physically or logically isolated resources from various networks. This new management function is not within the scope of a virtual network's capability, so it is necessary to introduce new manager or federation manager that acts as an interface or entry point of resource aggregation and sharing, and can be implemented as a centralized system controlling the interconnection of networks as a whole, or as a plug-in adaptation functions to each network. One of the use cases of federation would be the cloud computing. The computing power can be expanded by federating independent, heterogeneous clouds with the same virtual network management, which also leads to seamless deployment of application services over the networks.

10.5.1 Use Cases of Network Virtualization

In this section, the systematic knowledge of the typical use cases of network virtualization is described in order to clarify the purpose of network virtualization technology. The systematic description can be used as a guideline for describing the use cases.

10.5.1.1 Experimentation on the feasibility of new network architectures [5]

The main goal of this use case is to test the feasibility of the new network architectures in large-scale experimental networks or testbeds. The feasibility test of new technologies is essential for fostering innovation in network architectures. In order to perform feasibility test, researchers or developers will require a virtual network, which can work with customized protocols to test newly, developed research ideas. Many experiments can be performed in a single experimental networks or testbeds using virtualization technologies.

The experimental networks and testbeds provide the complete network environment and the operation experience of new network architectures and innovative network technologies. The experimental networks and testbeds also show key features of the systematic knowledge. Thus, experimentations over experimental networks or testbeds can be used to find unexpected effects and to overcome the limitations of the implementation.

10.5.1.2 Network virtualization for application service providers

This use case is related with application service providers who own the computer systems and local area network infrastructures. The application service provider enables their customers to use the computer systems. Application service providers deploy computer systems in multiple locations to improve their service availability. When the systems in location (A) are out of order, all applications running on these systems will be relocated to the systems in another site (B) instead. In this case, it is desirable that the application-user is not aware of this change. Thus, the systems deployed in different locations should be handled in the same manner as those located in the same location. One virtual computer system consists of multiple computer systems deployed in different locations, and one virtual network output link consists of multiple network output links deployed in different locations. The virtual computer system and virtual network output links should be synchronized.

10.5.1.3 Network virtualization for the network service provider

Today there are network service providers who own several different networks and provide several different services on each network. Networks should be operated using its management policy according to each service quality. For example, PSTN, dedicated line service, internet service, and mobile network service. Logical router technology is implemented in commercial products. With this technology, multiple virtual routers can be created on one router, and IP routing protocol can be performed on virtual routers independently. With programmable flow switching technology, the forwarding table of switch nodes can be controlled by the external servers. The management framework for virtual networks, which consist of multiple-layer resources (ex. WDM-circuit and packet-switching networks) are under development.

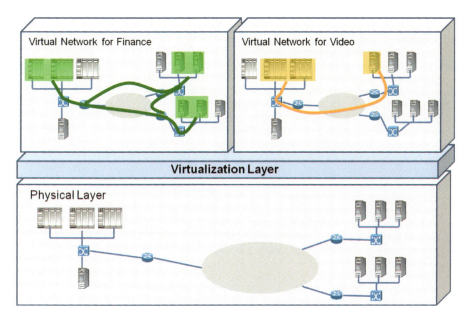

Fig. 10.2 Optimized virtual networks for service characteristics.

The virtual networks must be completely isolated from each other and the physical resource be optimally used.

10.6 Conclusions

This document describes the framework of network virtualization for Future Internet. It presents the concept and overview of network virtualization and virtual network that will be provided by network virtualization technology. This document also discusses the key challenges of supporting network virtualization over physical network infrastructure and investigates several requirements for realizing network virtualization. Finally, applicability and several use cases of network virtualization are also discussed.

References

[1] J. Rexform and C. Dovrolis, Point/Counterpoint future internet architecture: Clean-slate versus evolutionary research, *Communications of the ACM*, vol. 53, no. 9, September 2010.
[2] ITU-T Recommendation Y.3001 (2011), Future networks: Design goals and promising technologies.

[3] ITU-T Focus Group on Future Networks FG-FN OD73 (2010), Framework of network virtualization for future networks.
[4] S. Jeong and D. Colle, Virtual networks problem statement, draft-jeong-vnrg-virtual-networks-ps-00.txt (Work-in-progress), December 2010.
[5] GENI: Global Environment for Network Innovations GDD-06-08 (2006), GENI Design Principles.
[6] Browne, J. et al., "Classification of flexible manufacturing systems," The FMS Magazine, pp. 114–117, April 1984.

11

Interoperability, Standardisation and Governance in the Era of Internet of Things (IoT)

Vandana Rohokale[1], Rajeev Prasad[2], Dr. Neeli Prasad[1] and Prof. Ramjee Prasad[1]

[1]*CTIF, Aalborg University, Denmark*
[2]*GISFI, India*

11.1 Introduction

The idiom "Internet of Things" has appeared to describe a number of technologies and research disciplines that enable the Internet to reach out into the real world of physical objects. Technologies such as RFID, short-range wireless communications, real-time localization, and sensor networks are becoming increasingly common, bringing the "Internet of Things" into industrial, commercial, and domestic use. Since the last decade, a thought-provoking idea is fast emerging in the wireless communications which gives us the new scientific insight about the omnipresent presence around us of a variety of "things" or "objects," such as RFIDs, sensors, actuators, mobile phones, which, through unique addressing schemes, are able to interact with each other and cooperate with their neighboring "smart" components to reach common goals. This novel paradigm, named "The Internet of Things" (IoT) continues on the path set by the concept of smart environments and facilitates the way to the deployment of numerous applications with a significant impact on many fields of future everyday life. In this perspective, logistics, Intelligent Transportation Systems (ITS), business/process management, assisted living and e-health are only a few examples of possible application fields in which this novel paradigm will play a leading role in the near future.

The chapter is organised as follows. Section 11.1 describes the Interoperability in the future Internet in details. Standardisation for IoT is discussed in Section 11.2. Next, Section 11.3 addresses specific issues in IoT standardisation. Section 11.4 illustrates the security and privacy issues in the Internet of Things. Architecture and governance of IoT are taken into account in Section 11.5. Different network technologies for IoT scenarios are considered in Section 11.6. IoT middleware is elaborated in Section 11.7. Next Section 11.8 describes network technology. Section 11.9 enumerates Application areas and Industrial deployment. Section 11.10 gives insight about the socio economic ecosystems. Last Section 11.11 concludes the chapter with some focus on the technological challenges in the field of IoT.

11.2 Interoperability in the IoT

Interoperability is nothing but a guarantee of effective communication between people (personal computers, mobile phones, PDA and social networks, etc) and objects. In this sense, ITU has already identified RFID as the technology to be used to connect objects. The novel paradigm, named IoT is a fast emerging stimulating idea in the wireless scenario in which a variety of things or objects around us, such as RFID, sensors, actuators, mobile phones etc., are able to interact with each other and cooperate with their neighboring smart components to reach common goals. IoT is a world where objects can automatically communicate with each other and with the Internet providing services for the betterment of humanity. It is an environmentally aware Internet technology where objects do not necessarily have internet addresses. IoT supports many input-output devices and sensors like cameras, microphones, keyboards, speaker, displays, near-field communications (NFC), Bluetooth, accelerometers, etc. The main component of the IoT is the RFID system. RFID can automatically identify stationary or moving entities. An RFID system consists of RFID tags, readers and antennae. The structure of an RFID system is as shown in Figure 11.1. Each RFID tag has a unique electronic product code (EPC), which can be used to identify the object uniquely. A reader is used to read the data stored in RFID tags or to add new information to the tags. For transmission of radio frequency signals between the reader and tag, antennae are used [1].

11.2 Interoperability in the IoT

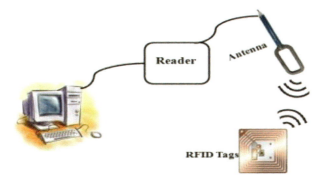

Fig. 11.1 RFID System.

Fig. 11.2 Conceptual Structure of IoT.

The main aim of IoT deployment is nothing other than to monitor and control objects via the Internet. The idea behind it consists of interconnecting objects by sensors and monitoring them via the Internet. Figure 11.2 elaborates the IoT concept. IoT seems to follow the famous Metcalf's law which states that the value and power of a network increases in proportion to the square of the number of nodes in the network. For the future Internet and IoT, it is very much essential to control and keep track of the immensely growing number of networked nodes so that it will be possible to network them in everyday devices in homes, offices, buildings, industries, transportation systems, etc. in cost-effective and valuable ways [2].

In the computing view, the IoT refers to a wireless and self configuring network among objects such as household applications as shown in Figure 11.2. The exponentially increasing quantity of objects around us needs proficient interaction schemes allowing anybody to easily access anything. Personal Area Networks (PAN) always refers to the IoT networks. A PAN is an unpredictable and spatially local network that includes every object a person could interact with. As with the IoT, the nodes of the PAN are the hardware-constrained devices. These nodes usually have a few kilobytes of volatile memory as well as non-volatile memory, a CPU with a few MHz frequency, limited energy, and communications via radio communications links such as Bluetooth, Zigbee, USB, Wi-Fi, etc. [3].

Pervasive computing environments in coordination with IoT is changing the paradigm of the association of one person with one networked computer, to a paradigm in which each person interacts with several networked objects. Coordination and cooperation is extremely important for the effectiveness, performance and quality of service (QoS) of composite systems. A careful incorporation of truly cooperative mechanisms enables an enrichment of the overall system behavior and contributes to progress in terms of flexibility, adaptibility, maintenance, inconspicuousness and service quality [4]. For identification of people and objects without any human intervention and to understand their specific functioning state, suitable technologies include RFID and complementary technologies such as sensor networks, GPS and Wi-Fi [5, 6].

There is a need for a standardized system for the identification of objects and the *Electronic Product Code (EPC)* is playing a vital role in this. In addition, a system to promote technology integration with seamless access directly from the application level would be very beneficial to speed up the development process. The development of RFID middleware to address these challenging items is a unique possibility to successfully ensure data interchange and technology integration [7]. The solution is known by the term middleware. This consists of one or more software layers placed between the technological and the application layers. Because of its most important role in simplifying the development of new services and the integration of legacy technologies into new ones, the middleware is receiving particular attention. Due to middleware, the programmer is relieved from needing to have detailed knowledge of the vast set of edge technologies adopted by the lower technology layers. Recently proposed middleware architectures for the IoT often follow the Service Oriented Architecture (SOA) approach. The horizontal layers of an enterprise system are depicted through the use of common interfaces and standard protocols. The process of designing the workflows of coordinated services is closely associated with object actions that are possible due to the appropriate development of business processes enabled by a SOA [8]. The beauty of the SOA approach is that it permits the reuse of the software and hardware platform, because a specific technology for the service implementation is not imposed by SOA [9].

Since a universally accepted layered concept is missing in SOA architecture, the proposed solution by [9] may face essentially the same problems of abstraction of device functionalities and communication capabilities, making provision of a common set of services and service composition environment. The middleware architecture for IoT based on SOA principles which addresses the middleware issues with a complete and integrated aspect is depicted in Figure 11.3. All the functionalities of the system are provided to the final user through the topmost application layer. It is not a part of middleware but exploits all the functionalities of the middleware layer. The service composition layer acts in a true sense as a repository of all presently connected service instances that are executed or invoked at run time to construct the composed services.

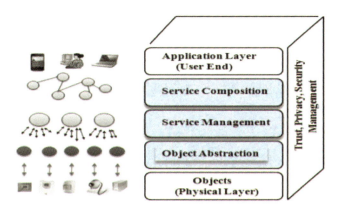

Fig. 11.3 IoT Middle Architecture based on SOA [9].

Different workflow standard languages include Business Process Execution Language (BPEL) and Business Process Modelling Language (BPML). Web Services Definition Language (WSDL) is typically used to describe the functional methods, input parameters and output values provided by web service interfaces within a SOA. Workflows can be nested in loops.

Certain kinds of semantic functionalities like policy and context management, QoS management and lock management are included in some middleware proposals for service management. Service management provides the major functionalities that are necessary for each object and that allow for their management. It also includes dynamic discovery of objects, status monitoring and service configuration [10]. Object abstraction is capable of harmonising the access to different devices through a common language and procedures. Since IoT consists of an enormous and heterogeneous set of objects, with each object providing specific functions accessible through its own language, there is a need for an abstraction layer which can take care of the harmonisation issue. Object abstraction acts as an interface sub-layer like a web-services interface to manage all messaging operations. It also functions like a communication sub-layer for the translation of the messages to object-specific commands. Lightweight implementations of the TCP/IP transport protocol stack such as TinyTCP, mIP and IwIP provide a socket like interface for embedded applications [11].

11.3 Standardisation For IoT

The research community is steadily contributing towards the full deployment and standardisation of IoT paradigm. Most of the efforts are contributed by Auto-ID Labs, European Commission, Standards Organisations such as ETSI, CEN, CENELEC, ISO, IETF, ITU and initiatives like GS1 EPCglobal, CASAGRAS, W3C etc. Table 11.1 summarizes different initiatives to facilitate standardization of IoT.

The aim of CASAGRAS (Coordination And Support Action for Global RFID-related Activities and Standardization) is to provide an incisive framework of foundation studies that can assist in influencing and accommodating international issues and developments concerning radio frequency identification (RFID) and the emerging Internet of Things, particularly with respect to standards and regulations.

11.3 Standardisation For IoT

Table 11.1. IoT initialization activities [9, 11, 12].

IoT Initiatives	Description	Objective
CASAGRAS	Coordination and Support Action for Global RFID related Activities and Standardisation-Embracing a fully inclusive range of EDGE technologies, including RFID for interfacing with the physical world	To provide an incisive framework of foundation studies that can assist in influencing and accommodating international issues and developments concerning radio frequency identification (RFID) and the emerging Internet of Things, particularly with respect to standards and regulations
GRIFS	Global RFID forum	To improve collaboration and thereby to maximise the global interoperability of RFID standards
RACE	Raising Awareness and Competitiveness in Europe for Networked RFID	Dedicated to contributing to a cohesive attitude and international cooperation about RFID
W3C	World Wide Web Consortium–Developing protocols to ensure long term growth for the web	The Semantic Web is an extension of the current Web that will allow you to find, share, and combine information more easily. Designed to be a universal medium for the exchange of data.
M2M	Cost Effective solutions for M2M Communication	M2M communications is the consumer demand for better quality of service (QoS), new applications, and increased mobility support.
ROLL	Routing Protocols for heterogeneous low power and lossy networks	Improvement in the overall Quality of Service for heterogeneous networks

CASAGRAS work packages include the following [13]:

- Standards and procedures for international standardizations in relation to RFID, including applications and conformance standards.
- Regulatory issues with respect to RFID standards.
- Global coding systems (GCS) in relation to RFID systems.
- RFID in relation to ubiquitous computing and networks.
- Functionalities including sensing, developments in RFID and associated standards.
- Areas of applications, existing and future, and associated standards.
- Socio-economic components of RFID usage.

W3C (World Wide Web Consortium) is involved in the following standardization efforts on IoT [15] :

- Mix of rapidly evolving networking technologies

- - Ethernet over twisted pair or coax
  - DSL over copper phone lines
  - Ethernet over building power wiring
  - WiFi and WiMAX
  - Bluetooth
  - ZigBee Sensor networks
- Cellular packet radio
- Challenges related to different addressing schemes in a P2P network
- Security and privacy issues in communication
  - Prevention of Phishing attacks, DoS attacks
  - Vulnerabilities of the Sandbox model of web browsers
  - Designing trust management solutions
- Device coordination for binding devices and services as part of distributed applications.
- Event transportation mechanisms — how to transport events to devices
  - Tunneling through NAT
  - Public and Private agents
  - Remote user interfaces
- Dynamic adaptation to user preferences, device capabilities and environmental conditions — server side and client-side adaptations based on policies

IoT Standardization Efforts in ANEC and BEUC [15]:

The safety of consumers is better ensured if product specific safety requirements are laid down and made directly applicable to the economic operators at the same time technical solution is considered as the responsibility of the standardisation forum. The European Consumer Voice in Standardisation (ANEC) and European Consumers' Organisation (BEUC) announced their inventory of consumer products containing nanomaterials. ANEC and BEUC started with monitoring the availability of products containing nanomaterials in 2009, and

Table 11.2. Technologies for IoT [9, 11, 12].

Wireless Technology	Communication Range (m)	Data Rate (kbps)
EPCglobal UHF Class 1 Gen 2 air interface protocol (ISO 18000-6C)	~1–10	Upto 640
IEEE 802.15.4	10–20	Upto 250
Bluetooth	10	Only 780
NFC	~10^{-2}	Up to 424
Wireless HART	10–100	~10^2
ZigBee	10–100	~10^2

that year's inventory listed only 151 products. The 2010 inventory includes 475 products in categories such as child products, food and drink, cosmetics, products for cars, and electronic devices. ANEC and BEUC support the Belgian Presidency's proposal to ensure the traceability of nanomaterials.

Standardization efforts are towards making standards that are open and have the following features:

- Interoperability
- Appropriate Identification for the involved entities
- Neutrality
- Trustworthy
- Transparent in governance
- Protects privacy and fundamental rights of users
- Security
- Liability and accountability
- New sections on health, safety, and environmental aspects of IoT are added.

11.4 Specific Issues in IoT Standardisation

EPC middleware is a software process and filter streams of tag data from multiple devices prior to sending to enterprise application systems as depicted in Figure 11.4. EPC middleware acts as an information router. EPC information system (EPCIS) is a primary vehicle for data exchange between subscribers; it encapsulates all access to underlying RFID infrastructure. EPCIS facilitates capturing; securing and access to EPC related data through a uniform interface.

Due to heterogeneous communication environment requirements for the IoT, vitality of Dynamic Spectrum Access (DSA) techniques are coming into

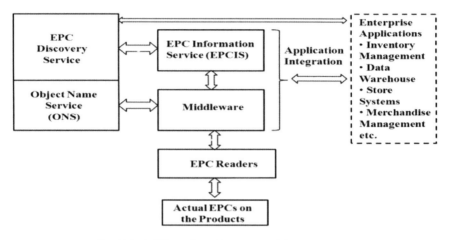

Fig. 11.4 EPC Network Architecture containing EPCIS [14].

the picture. Many devices share the same available spectrum by making use of autonomous network configuration from time to time. Standards regarding spectrum allocation, radiation power levels and communication protocols will ensure IoT cooperation with other radio spectrum users including message broadcasting, mobile telephony, disaster services, etc. Efficient implementation of Cognitive Radio Technology can be the proper solution for this. CIMIT (Center for Integration of Medicine and Innovative Technology) initiated a program in 2004 to lead the development of open standards for medical device interoperability, which supports the following systems [15].

- Clinical decision support systems
- Smart medical alarms
- Medical device safety interlocks
- Closed-loop control of medication delivery
- Remote healthcare delivery (home, battlefield, e-ICU etc.)
- Complete, accurate electronic medical records
- Hospital emergency preparedness
- Increase quality and completeness of research databases.

11.5 Security and Privacy Issues

Through IoT, the exchange of goods and services in global supply chain networks has an impact on the security and privacy of the involved stakeholders.

11.5 Security and Privacy Issues

Measures ensuring the architecture's resilience to attacks, data authentication, access control and client privacy need to be established [16]. The community will not accept the IoT unless their privacy is not adversely affected. The great amount of mistrust is towards the use of data collection by the IoT technology. The important question arising from IoT is about how much privacy the community is ready to relinquish in exchange for achieving a certain level of security. Security and privacy should not have an inversely proportional relationship; instead both of them should walk hand in hand with each other. IoT is looking for a security technology that will allow users to trust product lifecycle management intuitives that are normally hidden from the user. The key problem related with security is the authentication and data integrity. For resource-constrained RFID or sensor nodes, it is impossible to exchange the messages between the nodes more frequently.

There are two distinct approaches to security. In the symmetric or private key approach, which can be viewed as a child of the very early security attempts? For symmetric key approach, both sender and receiver possess a common secret key which is used for encryption as well as decryption. Distribution of such common secret keys and keeping them secret leads to vulnerabilities thereby needing new mechanisms. In asymmetric or public key cryptography, each user has a separate private key, which is known to the respective user only and a mathematically related public key which can be made public and freely distributed. Classification of security mechanisms is shown in Figure 11.5.

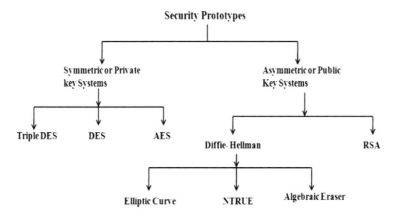

Fig. 11.5 Classifications of security mechanisms.

Table 11.3. Comparison of different Security Algoritms.

Algorithm	Encryption/Decryption	Digital Signature	Key Exchange
RSA	Yes	Yes	Yes
Diffie-Hellman	No	No	Yes
DSS	No	Yes	No
Elliptic Curve	Yes	Yes	Yes

Ensuring that the distributed public-keys are authentic is essential for the security of the system and a major security concern is known as the "man-in-the middle attack." In cooperative wireless communication, security against man-in-the-middle type of attacks is very much essential. Security of private key cryptosystems depends on the secrecy of the secret key. In the case of public key systems, it is infeasible to derive the private key from the public key. Breaking of a public key is a complex and timely task [17].

The use of public key cryptosystems can be broadly classified in three types as [18]:

- Encryption/decryption: The sender encrypts a message with the recipient's public key.
- Digital signature: The sender "signs" a message with the sender's private key. Signing is achieved by a cryptographic algorithm applied to the message or to a small block of data that is the result of applying a hash function to the message.
- Key exchange: Two sides cooperate to exchange a session key.

Existing sensor network solutions cannot be directly applied to the IoT scenario. For IoT, sensor nodes must be seen as nodes of internet; only then will it be possible for the nodes not belonging to the sensor network to authenticate the sensor network node. None of the existing security solutions can withstand the man-in-the-middle attack or proxy attack. The existing good security solutions needed to be modified according to the IoT perspective. For mobile nodes, the data can be kept secured while traversing through the network by making use of one-way hash message authentication mechanisms [19].

Currently, RFID solutions should be used in noncritical contexts: In the IoT, RFID can provide information about the presence of objects and possibly also data collected by sensors but system designers should address the risk of malicious alteration or interception of such information. Also, for RFID

11.5 Security and Privacy Issues

solutions to be integrated in the IoT, a special middleware must be used to provide a suitable interface, possibly exposing a web service interface to allow for remote interaction. In this case, services will of course be provided by a central architecture or by the middleware [20]. 13.56 MHz RFID-based Near Field Communication (NFC) is sometimes seen as an answer to some of the security issues of RFID. Mifare and NFC compliant devices provide authentication and symmetric ciphers, though it has been already demonstrated that reverse engineering is practicable and can compromise the entire system [21].

Privacy of communication and user data: A number of technologies have been developed to achieve information privacy goals. Some privacy enhancing technologies (PET) are:

- Integrating policy-based release of data
- Virtual Private Networks (VPNs): Impractical beyond the borders of the extranets.
- DNS Security Extensions (DNSSEC)
- Onion Routing
- Private Information Retrieval (PIR)
- Transport Layer Security (TLS): Transport Layer Security (TLS) and its predecessor, Secure Sockets Layer (SSL), are cryptographic protocols that provide secure communications on the Internet for such things as web browsing, e-mail, faxing, instant messaging and other data transfers. There are slight differences between SSL 3.0 and TLS 1.0, but the protocol remains substantially the same.

For privacy protection in IoT, [21] proposes secure multi-party computations. Many conventional problems involving scattered computations can be reformulated in a privacy-preserving manner. The secure multi-party computation creates new opportunities in the area of development of privacy-preserving ubiquitous applications. Secure multi-party computation techniques can be used to develop novel applications with privacy preservation as an essential objective. However, most of the secure multi-party computation protocols are inefficient and thus they are unsuitable for ubiquitous computing environments. They could be used to create privacy-preserving repositories containing both public and private information about subjects and objects that can be used to make decisions while protecting user privacy.

11.6 Architecture and Governance of the IoT

The architecture of IoT applications will require effectively enhanced service discovery strategies, tools and architectures that may in turn impact the overall business models and governance policies of the IoT. For the successful implementation of international network structures like IoT or simple internet, the guidelines from legitimate authorities play a vital role. There are certain challenges for the public governance of IoT mechanisms as mentioned below:

- Identification and addressing of the object
- Responsible authorities for identifiers
- Tracking and discovering object information
- Assurance of Information Security
- Ethical and Legal IoT Framework
- Overall control mechanisms

The concept of "multiple stakeholders in governance" is very important and it should be apparent as the new way forward, as a way to include the whole of society. This kind of advancement challenges the traditional legal and political understanding of legitimacy and makes it necessary to tackle the general question of who could be a legitimate stakeholder. Consequently, architectural principles are to be developed and compiled within an international legal framework; representation only has a legitimizing effect if the outcome reflects the values of the represented stakeholders. In particular, such a concept calls for procedures that establish equal bargaining powers and fair proceedings, as well as enhanced transparency and review mechanisms that enable the allocation of accountability. Intelligibility and unbiased access promote the mobilization of civil society and influence the architectural and legitimate principles of a government, such as flexibility and openness. The achievement of a greater degree of clarity and predictability also fosters the stability of a legal framework [22].

For the internet to act as a medium for mechanisms reaching towards equilibrium, it needs to:

- be free, open, uncensored, accessible, multilingual
- Support universal connectivity
  - Goal: allow bottom up, vertical and lateral information flows

11.6 Architecture and Governance of the IoT

- o Technology neutrality (un-tethered)
- o IPv4 and IPv6 must effectively cohabitate
- o Freedom of cross-border data flows.

Emerging patterns of ICT-enabled interaction that are transforming

- Economic, social and government structures (the information economy and society, etc.)
- Communication among individuals, groups and communities (social networking, web 2.0, etc.)
- Interactions between natural and artificial environments (the Internet of Things, ubiquitous networks, etc.) [23].

Figure 11.6 depicts the essential basic idea for architecture of Internet of Things. The transition from Intranet of things towards Internet of things is nicely depicted in the figure. The figure explains the evolution through engineering view.

Key objectives of IoT-A are mentioned below [24]:

- Create the architectural foundations of an interoperable Internet of Things as key dimension of the larger Future Internet
- Develop an architectural reference model together with an initial set of key building blocks

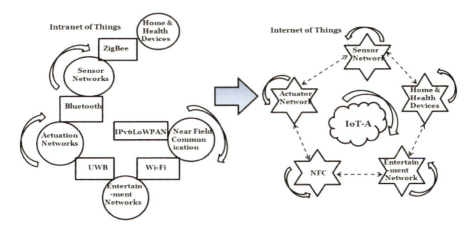

Fig. 11.6 Interoperable IoT architecture.
Source: European Lighthouse Integrated Project.

- Not re-inventing the wheel but federating already existing technologies
- Demonstrating the applicability in a set of use cases
- Removing the barriers to deployment and wide-scale acceptance of the IoT by establishing a strongly involved stakeholder group
- Federating heterogeneous IoT technologies into an interoperable IoT fabric.

Figure 11.7 depict the journey of IoT from RFID to a fully fledged Internet of Objects. In its childhood, IoT was perceived as only being linked to RFID. Later on, RFID tags were combined with some intelligence instead of being used merely as passive labels, which progresses towards the vision of the IoT. The second phase was truly dominated by sensor networks that have communication capabilities. This characteristic of WSNs was exploited for Near Field Communication (NFC) during the second phase. The third phase has developed two innovative aspects of IoT including Real World Internet (RWI) and Augmented Reality (AR), which are born from virtual reality technology [25]. This research work gives insight about the three important facets of future Internet architecture such as user-centric, object-centric and content-centric.

European commission project CORDIS FP7 takes into account each and every aspect of the future Internet architecture as shown in Figure 11.8.

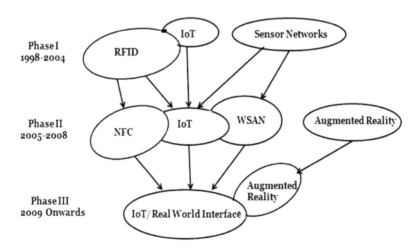

Fig. 11.7 Conceptual architecture of IoT evolution.

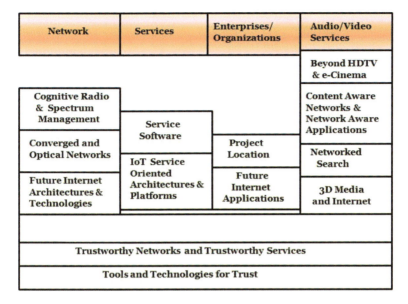

Fig. 11.8 Conceptual Architecture of Future Internet.
Source: European Commission, CORDIS FP7 project on Pervasive and trusted network and service Infrastructures).

Cognitive Radio networks can play a vital role in the future internet paradigm. Spectrum management with cognitive radio has the capability to resolve the entire spectrum conflicts evolved due to the multi-faceted nature of the IoT which includes many interfaces from various networks like ZigBee, Bluetooth, WPAN, WLAN, etc. [26].

11.7 Middleware or Software Platform for IoT

RFID and sensor networks and devices have great prospective and it is really a complex challenge to establish interaction between them. For RFID and sensor nodes to link seamlessly through a common platform named middleware is proposed by European consortium. Middleware is the active interface in between physical layer and actual user or application layer. Figure 11.9 shows the conceptual diagram of IoT Middleware.

Aspire

ASPIRE is an Advanced Sensors and lightweight Programmable middleware for Innovative RFID Enterprise applications. The ASPIRE middleware is being

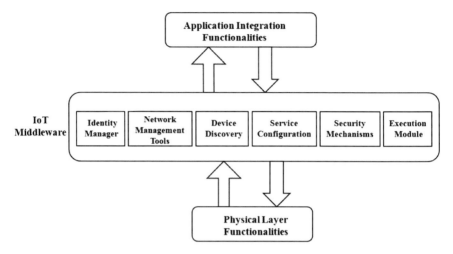

Fig. 11.9 IoT middleware or software platform.

placed at the heart of RFID infrastructures. In a Radio Frequency Identification (RFID) based scenario, the tags act as hosts for the resources in form of Electronic Product Codes (EPCs), IDs or other information as well as for value-added information in form of e.g., sensor data. The resource hosts are abstracted through the RFID readers due to the passive communication of the tags. The Object Naming Service (ONS) corresponds to the Entity Directory that returns the URLs of relevant resources for the EPC in question — this is the White Pages service. The EPC Information Service (EPCIS) implements the Resource Directory by storing more rich information of the resource. The ALE operates through an Event Cycle specification (ECspec) where resources are defined. A Business Event Generator (BEG) is introduced by ASPIRE which implements additional logic for interactions using semantics of the specific RFID application. The ASPIRE middleware is vendor and frequency independent [27].

Hydra

Hydra is a networked embedded system middleware for heterogeneous physical devices in a distributed architecture. Hydra aims at development of service oriented architecture (SOA) based middleware. Hydra middleware structure is proposed to have a transparent communication layer, similarly supporting centralized and distributed architectures. Security and privacy issues are taken into consideration while designing the Hydra middleware. It is expected to work well with resource constrained wireless nodes. Hydra is proposed to work on wired or wireless networks of distributed devices like wireless sensor networks or networks with RFID [28].

11.8 Network Technology

In fact, the "things" composing the IoT will be characterized by low resources in terms of both computation and energy capacity. Accordingly, the proposed solutions need to pay special attention to resource efficiency besides the obvious scalability problems. Medium access control and management continues to be a subject extensively investigated within the IoT arena, since it heavily affects the communication performance in terms of delays, losses, throughput, and reliability. When devising new strategies, it is important to focus on the self-managing, self-configuring, and self-regulating features, especially if the resulting network has to operate in emergency areas as in the case of the management of catastrophes [29]. The inevitable performance improvement due to cooperative wireless communication has gained the attention of many researchers towards cooperation between wireless networks. In particular, a novel architecture, namely Cellular Controlled Peer-to-Peer (CCP2P), has been introduced in the last years. According to it, cellular devices can create cooperative clusters with neighboring devices in their proximity using a short range technology. Each terminal is then contributing to the cooperative cluster by sharing its cellular link. The grouped members acting in a cooperative manner can achieve better performance than a standalone device, in different scenarios [30].

However, still great efforts are needed to make the cooperation effective and reliable in realistic scenarios where the peers are continuously moving and the channel is disturbed by concurrent communications in the same geographical area. Whereas cooperation can improve the performance in terms of energy consumption, broadcast techniques allow for reaching the widest area in the shortest possible time which is a feature highly demanded in case of emergency transmission. This is the case of safety-related applications in self-organizing Vehicular Ad-hoc NETwork (VANET), where vehicles share and distribute information by rebroadcasting a received information packet to their neighbors. An efficient broadcast technique can offer a high reactivity without sacrificing the communication reliability [31].

When dealing with multimedia communications, flow control is the critical issue for heterogonous wireless networks used in the IoT scenario. The work in paper [32] focuses on the design of control policies and proposes an approach based on the maximization of the average throughput over the wireless last-hop, under constraints on the maximum connection bandwidth allowed at the application layer, the queue-capacity available at the Data-Link layer, and the average and peak transmit energies sustained by the Physical layer. Paper [33] addresses the issues of interconnecting geodata sensor networks when deployed to monitor environmental factors in an extended geographical area. Issues related to the bandwidth management and security, when short messages are transmitted through the backbone network is considered in the IoT context which makes use of MPLS protocol.

In Cognitive Radio Network, cognitive radio adapts its services according to the changes in its surrounding, due to which, spectrum sensing has become an important requirement for the realization of CR networks [34]. The major necessary functionalities for spectrum sensing in cognitive radio ad hoc networks are as follows:

- Primary User (PU) Detection: the CR node continuously monitors and analyzes its local radio environment, for determining holes in the spectrum.
- Cooperation: the observed information in each CR node is exchanged with its neighboring network nodes for the improvement in sensing accuracy in case of hidden node problems.
- Sensing control: it enables each CR node to perform its sensing operations adaptively to the dynamic radio environment. Also, it coordinates the sensing operations of the CR nodes and its neighbors in a distributed manner, which prevents false alarms in cooperative sensing.

The cooperative wireless communication (CWC) concept is more applicable to wireless sensor networks and cognitive adhoc networks than that of cellular networks. In cooperative communication, the information overheard by neighbouring nodes is intelligently used to provide the healthy communication between a source and the destination (called a sink). In CWC, several nodes work together to form a virtual array. The overheard information by each neighboring node or relay is transmitted towards the sink concurrently. The cooperation from the network nodes that otherwise do not directly contribute in the transmission is intelligently utilized in CWC. The sink node or destination receives numerous editions of the message from the source, and relay(s) and it estimates these inputs to obtain the transmitted data reliably with higher data rates [35].

For improvement in the QoS of IoT, a cooperative approach may be applied with the cognitive radio networks. There are two main cooperative approaches towards cognitive radio (CR) viz. Commons Model and Property Rights Model. In the commons model, primary terminals are unaware of the presence of secondary users, thus behaving as if no secondary activity was present. Instead, secondary users sense the radio environment for finding out spectrum holes and then take advantage of the detected transmission opportunities. For the property rights model, the primary nodes may accept to lease their bandwidth for a fraction of time; in exchange for the concession, they benefit from the superior QoS in terms of rate of outage probability and improvement in energy savings. Cooperative relaying of primary packets through secondary nodes has been proven to be a promising technique to improve the secondary throughput by utilizing the idle periods of the primary nodes [36].

A summary of four IEEE based networking protocols and their parametric comparison is elaborated in the Table 11.4. Among all of them, Bluetooth and ZigBee are more suitable for the low data rate applications with constrained resources such as mobile adhoc devices and battery operated sensor networks, which results in increased lifetime of such miniature devices. For high data rate counterparts, such as audio/video surveillance systems, UWB and Wi-Fi may prove to be a better solution because of their higher efficiency results [37].

11.9 Application Areas and Industrial Deployment

As the technologies needed for the Internet of Things become available, a wide range of applications will be developed. These can support policy in areas including transportation, environment, energy efficiency and health. Huge benefits will come not only from faster productivity growth, but also

Table 11.4. Parametric Comparison of Wireless Networking Protocols.

Standard	Bluetooth	UWB	ZigBee	Wi-Fi
IEEE spec.	802.15.1	802.15.3a	802.15.4	802.11 a/b/g
Frequency band	2.4 GHz	3.1–10.6 GHz	868/915 MHz: 2.4 GHz	2.4 GHz; 5 GHz
Max signal rate	1 Mb/s	110 Mb/s	250 Kb/s	54 Mb/s
Nominal range	10 m	10 m	10–100 m	100 m
Nominal TX power	0–10 dBm	−41.3 dBm/MHz	(−25)–0 dBm	15–20 dBm
Number of RF channels	79	(1–15)	1/10: 16	14 (2.4 GHz)
Channel bandwidth	1 MHz	500 MHz–7.5 GHz	0.3/0.6 MHz; 2 MHz	22 MHz
Modulation type	GFSK	BPSK, QPSK	BPSK (+ASK), O-QP	BPSK, QPSK, COFDM, CCK, M-QAM
Spreading	FHSS	DS-UWB, MB-OFDM	DSSS	DSSS, CCK, OFDM
Coexistence mechanism	Adaptive freq. hopping	Adaptive freq. hopping	Dynamic freq.selection	Dynamic freq. selection, transmit power control
Basic cell	Piconet	Piconet	Star	BSS
Extension of the basic cell	Scatternet	Peer-to-peer	Cluster tree. Mesh	ESS
Max number of cell nodes	8	8	>65000	2007
Encryption	E0 stream cipher	AES block cipher (CTR, counter mode)	AES block cipher (CTR. counter mode)	RC4 stream cipher (WEP), AES block cipher
Authentication	Shared secret	CBC-MAC (CCM)	CBC-MAC (ext. of CCM)	WPA2 (802.111)
Data protection	16-bit CRC	32-bit CRC	16-bit CRC	32-bit CRC

in many other ways like increasing efficiency in material handling and general logistics, efficiency in warehousing, product tracking, efficiency in data management, reducing production and handling costs, speeding the flow of assets, anti-theft and quicker recovery of stolen items, addressing counterfeiting, reducing mistakes in manufacture, immediate recall of defective products, more efficient recycling and waste management, achieving CO_2 reductions, energy efficiency, improved security of prescription medicine, and improved food safety and quality.

11.9 Application Areas and Industrial Deployment

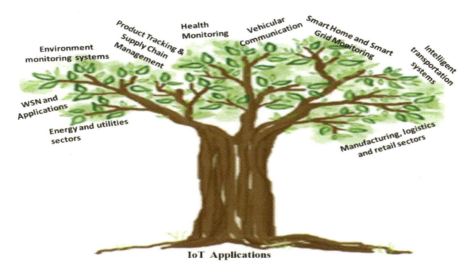

Fig. 11.10 Applications of Internet of Things.

Major application areas include the following as shown in Figure 11.10 [37, 38]:

- **Manufacturing, Logistics and Retail sector** has already seen a lot of RFID deployments and will continue to be a major user of RFID, wireless sensors and smart objects. Applications include product authentication and anti-counterfeiting, next generation industrial automation and supply chain management, inventory management, track & trace, and remote maintenance, service & support.
- **Energy and Utilities sector** include Smart electricity and water transmission grids, real time monitoring of water supply and sewerage systems etc. In addition to the utility infrastructure monitoring, there are many upcoming applications at the consumer end. Efficient energy and water consumption at homes enabled by devices connected to the grid are a major upcoming application area.
- **Intelligent Transportation Systems** applications include support for vehicular ecosystems, use of in-vehicle sensor networks, telematics, GPS and wireless networks for developing smart vehicles and transportation systems. Vehicle to vehicle communications and vehicle to roadside communications can be used for collaborative

road safety and efficiency. Vehicle tracking, traffic data collection for management, traffic rule enforcement systems, automotive infotainment systems are going to be a part of an integrated network.
- **Environment monitoring applications** are one of the most talked about applications of sensor networks. This will involve use of wireless sensor nodes for monitoring of weather, environment, civil structures, soil conditions, etc. and will have extensive use in agriculture, weather monitoring, security and surveillance, disaster management etc.
- **Home management and monitoring** applications will involve use of wireless sensor nodes, smart appliances, wireless networks, home gateways and internet for applications as wide ranging as home security, care of the elderly, smart energy control etc.

There are several other such applications already in use that impact society and the economy and it is expected that potentially many more will evolve as various IoT related technologies mature. This novel paradigm, named "The Internet of Things" (IoT) continues on the path set by the concept of smart environments and paves the way to the deployment of numerous applications with a significant impact on many fields of future everyday life. In this context, logistics, Intelligent Transportation Systems (ITS), business/process management, assisted living, and e-health are only a few examples of possible application fields in which this novel paradigm will play a leading role in the near future [39]. Some of the interesting futuristic applications of IoT may include Robotic Taxis, City Information Model, Enhanced Game Room, etc. as mentioned in the SENSEI FP7 EU project.

Benefits from IoT

- Real time monitoring is possible for efficient decision making and necessary action to be taken.
- It provides better transparency of physical flows and detailed state information which is important for regulatory compliance and public dissemination.
- Improved performance, visibility and scalability of business process automation.

- Creation or transformation of new and existing business processes by enhancing efficiency, accuracy, mobility and automation.
- Utilizes software control for initiation of sequence of events.

11.10 Socio-economic Ecosystems

Zhenjiang government of Beijing China in its press release has put forth a proposal for creation of a Hi-tech city through which the city will convert itself into an innovative Future Internet ecosystem. It will support remote healthcare, intelligent transport system and bring in different e-energy solutions for the betterment of mankind [40].

The future Internet will be built with the help of resource constrained devices such as sensor nodes and RFID. The socio-economic ecosystem solutions are going to play a vital role for this scenario. The paper [42] outlines the energy challenge and presents the role of education as a vector in changing behaviors related to energy consumption by adopting a sustainable attitude aimed at environmental protection. The emphasis is put on teaching activities that will enhance the capabilities of youngsters and teachers from secondary schools concerning the efficient use of energy and the promotion of Renewable Energy Sources. The role played by schools in a green world is strengthened by on-line training delivered to youngsters via e-books and portals specially designed for this purpose or by learning by doing via interactive games.

Recently, the energy consumption by data centers is increasing with a high pace. The computer science community concentrates its efforts towards reducing these electricity costs, while keeping performance at high level. The authors argue that the ecological footprint is not only related to the raw energy consumption but also depends on the production means of the electricity, some being obviously greener than others and that the full life-cycle of ICT equipment should also be considered. [42]. Better-quality eminence is required for integrating environmentally sound choices into supply chain management research and practice. Green supply chain management has developed into an important and growing area of strategic advantage for many countries and companies. Closed-loop supply chains are being used to recover assets that would otherwise be lost. The research work in [43] investigates the performance and satisfaction regarding closed-loop supply chains in Cyprus. The findings and interpretations are summarized, the main research issues and opportunities are highlighted, and recommendations are made for improvements.

Paper [44] investigates the potential of smart routing in minimization of radio environment pollution in cognitive ad-hoc wireless networks. A wireless ad-hoc sensor network based on the IEEE 802.11b standard is used as a simulation testbed to determine statistical distribution of interference levels. Three different routing methods are used — flood routing, location aware routing (LAR) and a simultaneous localization and radio environment mapping based routing (SLAM routing). The authors of [45] propose a Hybrid Energy Storage System (HESS) that combines a super capacitor with a rechargeable battery. They claim that by using the HESS flexible energy-aware cost-benefit function, significant extension of the network lifetime is achieved by means of a balance between the energy consumption and the reliable delivery of data packets.

The authors of [46] claim that they have built a fully functional prototype home appliance with a socially aware interface to indicate the comprehensive usage of the user's peer group according to these principles. This research work elaborates how home appliances might be enhanced to develop awareness of energy usage. The prototype designed in this case is based on a design of immediate feedback of aggregated information.

11.11 Conclusions and Future Work

To bring ubiquitous computing that is Internet of Things (IoT) into reality, there are many challenges like convergence of different services and functionalities, standardisation, Identification, security, Augmented reality, etc. This chapter is an attempt to address these infrastructural issues and possible ways to achieve it. Still there are many important issues to be addressed like powerful security solutions, generic middleware support where researchers are encouraged to focus on.

Acknowledgment

This work has been carried out in the scope of the ASPIRE project, funded by the European Commission in the scope of the FP7 programme under contract 215417. Authors would like to extend thanks and appreciation to colleagues in CASAGRAS-2, CTIF and GISFI.

References

[1] H. Y. Chen and C. H.Chen, "Mutual authentication protocol for RFID conforming to EPC Class 1 Generation 2 standards," *Computer Standards and Interfaces*, vol. 29, no. 2, 2007, pp. 254–259.

[2] R. A. Dolin, "Deploying the Internet of Things," *IEEE Computer Society, Proceedings of the 2005 symposium on Applications and the Internet.*

[3] S. Duquennoy, G. Grimaud and J.-J. Vandewalle, "The Web of Things: Interconnecting Devices with high usability and performance," *IEEE Computer Society*, DOI 10.1109/icess.2009.13.

[4] Manfred Bortenschlager, "Current developments and future challenges of coordination in pervasive environments," *doi.ieeecomputersociety.org/10.1109/WETICE.2007.132*.

[5] N. Kong, X. Li and B. Yan, "A model supporting any product code standard for resource adressing in the internet of things," in *Intelligent Networks and Intelligent Systems, 2008. ICINIS '08*, First International Workshop on 1–3 Nov. pp. 233–238.

[6] M. P. Michael, "Architectural solutions for mobile RFID services for the internet of things, In: *Congress on Services — Part I, 2008. Services'08*, IEEE 6–11 July, pp. 71–74.

[7] D. Bruneo, A. Puliafito, M. Scarpa and A. Zaia, "Mobile middleware and its integration in enterprise systems," Published on *Handbook of Entreprise integration, Mostafa Hashem Sherif*, 3rd edn., Taylor & Francis Chapter 6, New York, 10016, USA.

[8] L. Atzori, A. Iera and G. Morabito, "The Internet of Things: A Survey," *Elsevier Journal of Computer Networks* 54 (2010) 2787–2805.

[9] J. Pasley, "How BPEL and SOA are changing web services development," *IEEE Internet Computing* 9(3) (2005) 60–67.

[10] Hydra Midlleware Project, FP6 European Project, <http://www.hydramiddleware.eu>.

[11] S. Duquennoy, G. Grimaud and J.-J. Vandewalle, The web of things: Interconnecting devices with high usability and performance, *Proceedings of ICESS '09*, HangZhou, Zhejiang, China, May 2009.

[12] D. Giusto, A. Iera, G. Morabito and L. Atzori, "The Internet of Things," *20th Tyrrhenian Workshop on Digital Communications*, 2009.

[13] EU RFID cluster Project, "CASAGRAS — Coordination and Support Action for Global RFID related Activities and Standardisation," 01 Jan 2008.

[14] C. Balkesen, "An Internet of Things Infrastructure," GS1, EPC Global PC, March 2008.

[15] J. Sen, "Internet of Things — A Standardisation Perspective," invited presentation in *2nd GISFI (Global ICT Standardisation Forum of India) Standardisation Workshop*, New Delhi, June 21–23, 2010 India.

[16] R. H. Weber," Internet of Things — New security and privacy challenges," *Computer Law and Security Review* 26 (2010) 23–30.

[17] SecureRF White paper, "An Introduction to Cryptographic Security Methods and Their Role in Securing Low-Resource Computing Devices," May 2010.

[18] I. Anshel, M. Anshel, D. Goldfeld and S. Lemieux, "Key agreement, The algebraic eraser and light weight cryptography," *International workshop on Algebraic Methods in Cryptography*, November 2005.

[19] H. Krawczyk, M. Bellare and R. Canetti, HMAC: Keyed-Hashing for Message Authentication, IETF RFC 2104, February 1997.

[20] C. M. Medaglia and A. Serbanati, "An overview of privacy and security issues in the Internet of Things," *The Internet of Things, 20th Tyrrhenian Workshop on Digital Communications, Springer Science+Business Media*, LLC 2010.

[21] F. D. Garcia, G. de Koning Gans, R. Muijrers, P. van Rossum, R. Verdult, R. W. Schreur and B. Jacobs (2008) Dismantling MIFARE classic, *Proceedings of ESORICS 2008*, Malaga, Spain, pp. 97–114.
[22] J. Sen, "Privacy preservation technologies in Internet of Things," in *Proceedings of the Interntaional Conference on Emerging Trends in Mathematics, Technology, and Management*, pp. 496–504, Kolkata, January 11–12, 2011, India.
[23] R. H. Weber, "Internet of Things — Need for a new legal environment?" *Computer Law and Security Review* 25 (2009) 522–527.
[24] T. Vetter, "The internet of things: A killer application for global environmental sustainability?" IGF, Hyderabad, Dec. 2008.
[25] T. R. Tronco, T. Tome and C. E. Rothenberg, "Scenarios of evolution for a future internet architecture," T. Tronco (Ed.): *New Network Architectures*, Springer-Verlag Berlin Heidelberg 2010, SCI 297, pp. 57–77.
[26] European Commission, CORDIS:FP7 ICT projects, "Internet of Things: Architecture," 2010.
[27] Advanced Sensors and lightweight Programmable middleware for Innovative RFID Enterprise applications (ASPIRE), FP7, http://www.fp7-aspire.eu/.
[28] Networked Embedded System Middleware for Heterogeneous Physical Devices in a Distributed Architecture (Hydra), FP7, www.hydramiddleware.eu.
[29] E. Cipollone, F. Cuomo and A. Abbagnale, "A Distributed Procedure for IEEE 802.15.4 PAN Coordinator Electionin Emergency Scenarios."
[30] E. Scuderi, R. Parrinello, D. Izal, G. P. Perrucci, F. Fitzek, S. Palazzo and A. Molinaro, in "A Mobile Platform for Measurements in Dynamic Topology Wirele Networks."
[31] S. Busanelli, G. Ferrari and S. Panichpapiboon, "Cluster-based Irresponsible Forwarding."
[32] E. Baccarelli, M. Biagi, N. Cordeschi, T. Patriarca and V. Polli, "Optimal Cross-Layer Flow-Control for Wireless Maximum-Throughput Delivery of VBR Media Contents."
[33] M. L. Lobina and T. Onali, "A Secure MPLS VPN Infrastructure for Complex Geodata Sensor Network."
[34] I. F. Akyildiz, W.-Y. Lee, M. C. Vuran and S. Mohanty, "Next generation/dynamic spectrum access/cognitive radio wireless networks: A survey" I. F. Akyildiz et al./*Computer Networks* 50 (2006) 2127–2159.
[35] P. Liu, Z. Tao, Z. Lin, E. Erkip and S. Panwa, "Cooperative wireless communications: A cross-layer approach" *IEEE Wireless Communications* August 2006.
[36] V. Rohoakale, N. Kulkarni, H. Cornean and N. Prasad, "Cooperative opportunistic large array approach for cognitive radio networks," *8th IEEE International Conference on Communications*, June 2010, Bucharest, Romania.
[37] R. Prasad (ed.), P. Balamurlidhar, S. Bothe and P. Mishra, "Internet of Things (IoT) in Future Trends and Challenges for ICT Standardization, 193–227, 2010 River Publishers.
[38] Disruptive Technologies: Global Trends 2025, SRI Consulting Business Intelligence, Appendix F: The Internet of Things, 2008.
[39] J. Buckley, "From RFID to the Internet of Things," *Pervasive networked systems, Report on Conference organized by DG Information Society and Media, Networks and Communication Technologies Directorate*, 6 and 7 March 2006, CCAB, Brussels.
[40] Press Release of "Zhenjiang government, Nokia Siemens Networks to create hi-tech city," Machine-to-Machine communications platform to enable socio-economic development, Beijing, China, October 11, 2010.

[41] D. Banciul and A. Alexandru, "Greening dimension of learning in secondary schools," *Journal of Green Engineering*, River Publishers, vol. 1 no. 2, January 2011.
[42] J.-M. Pierson, "Green task allocation taking into account the ecological impact of task allocation in clusters and clouds," *Journal of Green Engineering*, River Publishers, vol. 1 no. 2, January 2011.
[43] S. Louca, A. Kokkinaki, "Closed-loop supply chains in ICT: Best practices and challenges in Cyprus," *Journal of Green Engineering*, River Publishers, vol. 1 no. 2, January 2011.
[44] D. Zrno, D. Šimunić, R. Prasad," Optimizing cognitive ad-hoc wireless networks for green communications," *Journal of Green Engineering*, River Publishers, vol. 1 no. 2, January 2011.
[45] N. Pais, B. K. Cetin, N. Pratas, F. J. Velez, N. R. Prasad, R. Prasad, "Cost-benefit aware routing protocol for wireless sensor networks with hybrid energy storage system," *Journal of Green Engineering*, River Publishers, vol. 1 no. 2, January 2011.
[46] Karlgren, L. E. Fahlen, A. Wallberg, P. Hannson, O. Stahl, C. Floerkernier et al. (eds.), "Socially Intelligent Interfaces for Increased Energy Awareness in the Home," IoT 2008, LNCS 4952, pp. 263–275, Springer Verlag Berlin Heidelberg 2008.

12

IoT Validation and Interoperability IoT — IPv6 and M2M

Marylin Arndt[1], Philippe Cousin[2], Patrick Grossetete[3], Latif Ladid[4], and Sébastien Ziegler[5]

[1]*Orange, France*
[2]*easy global market, France*
[3]*Cisco Systems, Inc., France*
[4]*SnT University of Luxembourg; President, IPv6 FORUM, Luxembourg*
[5]*MANDAT International, Switzerland*

12.1 M2M: A Set of Enabling Technologies Paving the Way From the Internet of Services Towards the Internet of Things

The so called "Internet of Services" is represented by a network and service infrastructure with dynamic resources. This infrastructure has self management capabilities and uniform service interfaces, allowing a kind of "plug and play" of Connected Machines or Devices, in the context of user services. The development of the Internet of Services and its enablers in terms of functionalities is often said to be one pillar of the more Global "Internet of Things."

M2M services and applications architecture describe a simple representation of the Internet of Services, as can be seen in Figure 12.1.

After a short historical recall of M2M story, the paper will firstly describe the state of the art standardization activity at the moment, considering the Internet of Services actual development. Then, a set of enabling technologies will be described.

In a second part, considering the remaining issues and the open challenges, the paper will go through the requirements not fully taken into account today for the Internet of Things Development and conclude with a global vision of the future.

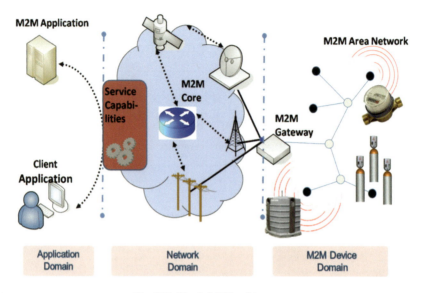

Fig. 12.1 Simple M2M architecture.

12.1.1 Short History in Time

M2M standardization became a strategic topic at ETSI [1] in 2008. M2M meant "*Machine-to-Machine*" but was not limited to only direct Machine to Machine communications; a better wording would have probably been "Machine Oriented Communications and Services."

The first commercial applications have focused on Vending Machines, Fleet Management, and Transport Applications, where the main concern was to establish connectivity between a set of machines and their remote business database. The mobile network was often used as the underlying network infrastructure because it was flexible and more securely designed. Rapidly the Industry faced an important issue of non-interoperability, either interworking, between these developments made in vertical silos. Therefore the need of standardisation on the service layer level, defining three main interfaces, of the service platform towards the Devices (mId), towards the Applications (mIa), and between the Devices and the Gateway (dIa). (See also Figure 12.4).

In 2009, the new technical committee M2M began to work at ETSI, and made the choice to specify the concept of a generic "horizontal" platform of services applicable to any vertical M2M application, with resource based interfaces (RESTful style).

Detailed objectives of the committee were:

- To collect and specify M2M requirements from relevant stakeholders;
- To develop and maintain an end-to-end overall high level architecture for M2M;
- To identify gaps where existing standards do not fulfil the requirements and provide specifications and standards to fill these gaps, where existing standards bodies or groups are unable to do so;
- To co-ordinate ETSI's M2M activity with that of other standardization groups and fora.

The workshops organized throughout the last three years showed another important movement into the industrial actors of M2M: The first newcomers were electronic modules and card manufacturers, then replaced (in number) by systems manufacturers, and then by network operators accompanied by ICT companies, showing the shift of the standardization needs from the lower to the higher layers. In the same period of time there was a multitude of various standardization initiatives on M2M that were held by Standards Developing Organizations (SDOs) such as ITU-T, ISO/IEC JTC 1, ETSI, IETF, IEEE, 3GPP, OMA, etc. The increasing number of participating companies to ETSI initiative, showed that a global coordination was considered as a necessary step.

And now, at the beginning of 2011, when the worldwide industry has crossed a difficult period of crisis and at the same time when Energy Efficiency and Sustainable Development became Key and added momentum and scale, wireless technologies for managing ad-hoc capillary networks became also mature. And now that all the devices can be connected at low cost, arises the opportunity of "Business driven Applications" to become "Mass Market driven." The generic approach taken by ETSI TC M2M now has a chance to go over the vertical preexisting developments, and this is one necessary step towards the Internet of Services Vision.

Figure 12.2 shows the expected evolution into the infrastructure brought by M2M standardization work.

12.1.2 M2M Services Standardization Framework Description

The scope of the work done at ETSI TC M2M was to develop standards as to support M2M architecture through a set of M2M Services Capabilities.

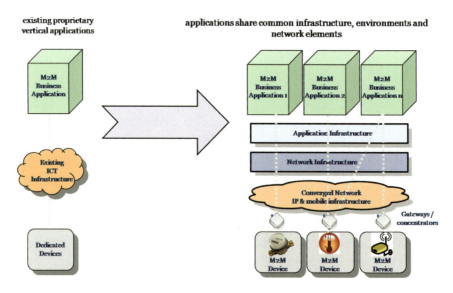

Fig. 12.2 Inverting the pipes.
Source: Helwett Packard.

After a first description of the global architecture with the main interfaces the set of service capabilities deduced from the Functional Requirements will be presented, ending with the definition of the APIs.

12.1.2.1 M2M high level system architecture

The following Figure 12.3 provides a High-Level M2M System Architecture view:

The M2M system Architecture includes an M2M Device domain, and a Network and Applications domain. The High Level System Architecture is based on existing standards regarding the network domain extended with M2M capabilities.

Mia, mId, and dIa interfaces are underlined into Figure 12.4.

12.1.2.2 M2M system requirements and basic service capabilities

The scope of the work done in ETSI TC M2M was to define a set of Service Capabilities, which will permit to build ontology for service development, including the specificities of Machine Oriented Communications. The System

12.1 M2M: A Set of Enabling Technologies Paving

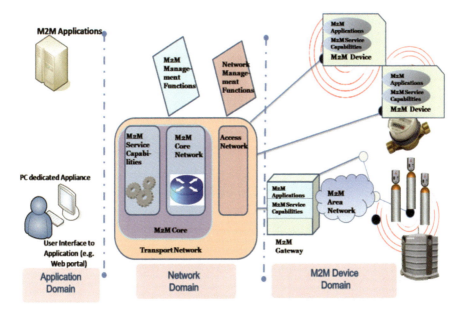

Fig. 12.3 High level system overview.

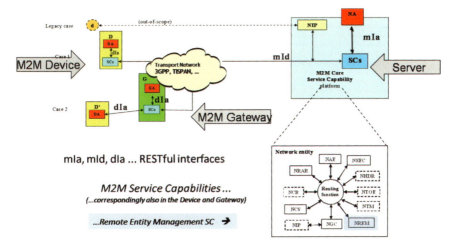

Fig. 12.4 Elements of M2M architecture.

is built with a "service centric" approach, which means that Capabilities will be derived from the Requirements, at the service level and on the network side.

The main important basic service capabilities are the following:

- **Connectivity Service Capabilities**

The M2M System should be able to establish a communication session for M2M Applications between the Network and Applications Domain and the M2M Device or M2M Gateway regardless of network addressing mechanism used, e.g., in case of an IP based network the session establishment should be possible when IP static or dynamic addressing are used.

The M2M services may support multiple communication means, e.g., SMS, GPRS and IP Access.

- **Machine oriented specific Service Capabilities**

  One important characteristic of the Connected Machines is the following: There is a very large diversity of "devices" that can connect to the network and their intrinsic capabilities can go from status "always on and connected" to "sporadically connected" and "sleeping." These properties must be taken into account by the service layer. For example, it should be possible to send a message addressed to a sleeping device. The M2M System will deliver the message to the device when it becomes awake.

  Another important thing is that due to the historical developments made up to now, the physical architecture is often composed of a set of devices connected and communicating behind a gateway. That implies that the M2M System should be able to address the M2M Devices or M2M Gateways using M2M Device Names or M2M Gateway Names respectively. This implies also that the M2M Gateway may be capable of interfacing to various M2M Area Network technologies.

- **Data Models and links with Business Databases and Applications, Data collection & reporting**

  The M2M System shall support the reporting from a specific M2M Device or M2M Gateway or group of M2M Devices or group of M2M Gateways in the ways requested by the M2M Application. As data models have been historically developed for vertical dedicated applications, a very wide variety of different data models exist today, and an intermediate representation using a meta_model will be useful to make the mIa interface generic.

  One important new aspect is that the M2M System will support mechanisms to allow M2M Applications to discover M2M

Capabilities offered to them, and this is one of the first "advanced" service capabilities that the system has to include and to standardize. Additionally the M2M device and M2M gateway shall support mechanisms to allow the registration of its M2M capabilities to the M2M system. These two functionalities built up over the basic set of service capabilities belong to the set of advanced service capabilities.

12.1.2.3 M2M system requirements and advanced service capabilities

As described just above, the M2M system description has allowed us to define the advanced Service Capabilities needed by the M2M System. Discovery and Registration Capabilities have been described as needed functionalities that must be available into the system. Another one concerns the need of remote Device Management, and this one has an essential feature, to be able to manage billions of machines firmware during their lifecycle (example: A meter has a 15 years life cycle!). Existing standards and methods already developed for other purposes (for example: Boxes management, Mobile phones management) will be reused.

Provisional mechanisms shall also be supported by M2M systems as to perform simple and scalable provisioning of M2M Devices, the devices being present or absent on the network.

The M2M Application or Capabilities in the Service Capabilities shall support auto configuration, without human intervention, when they are turned-on, and again with the need of registration to the M2M Application functions at this stage.

All the actors playing a role into the M2M development agree on one requirement which is considered as essential: Security. Security requirements are defined at all the levels of machine management functions, throughout all the underlying networks, and in an End to End service delivery vision. To add some details to the picture, Secure Applications (often called Trusted Applications) will include:

- Authentication procedures
- Authorization mechanisms
- Device integrity check and validation

- Privacy protection
- Security credential and software upgrade requirements

Identification, including Device identification, Naming, Numbering and Addressing functions, are functionalities that will be necessarily implemented on the platform.

As an example, the M2M System shall allow flexible addressing schemes, including:

- IP address of CO
- IP address of a group of COs
- E.164 addresses of CO (e.g., MSISDN)
- IMSI (International Mobile Subscription Identity)
- URL, e.g., SIP or e-mail addresses

12.1.2.4 Open APIs

Service providers will be able to deploy an application development platform which exposes the standard set of service capability features via the standard APIs as well as a standard stack for communications, power management and sensing, provided by the operators' M2M platform. These APIs can be exposed (or not) on the network to allow third parties to develop services using the APIs.

12.1.3 M2M Applications

Since the beginning of the M2M story, the market opportunities have grown ad were driven by vertical applications, very often business-oriented. Today the developments of platforms are pushed by a societal phenomenon called Energy efficiency and by the fact that populations are ageing, with an increased need of care (health, assistance, surveys...) in most advanced countries. Another trend concerns the profound changes that occur with the income of the Electric Vehicle into the City and into the everyday life. Last, but not least, with the worldwide success of the machine commonly known as the "mobile phone", one can imagine today that thousands of services can be delivered to the Mobile User, his Smartphone being connected everywhere to sensors at home through a M2M box and offering a portal to remotely manage his home consumption.

12.1 M2M: A Set of Enabling Technologies Paving 295

All of these applications are the target of huge investments in all segments of the Industry belonging to Electronic, Information, and Telecommunications, with new actors always looking for opening new market domains.

In ETSI TC M2M, five Use Cases Work Items have been opened, corresponding to these domains, as to set up specific requirements and to check if they are taken into account within the System View and into the architecture of the "horizontal" platform. These Use Cases are:

- Smart Metering
- Smart City
- Smart grid
- E-Health
- Automotive
- Connected Consumer

It is not the hazard if quite all these use cases are or will be the target of a European standardisation Mandate. Standardisation mandates are tools used by the European commission and sent to their agreed SDOs, CEN, CENELEC and ETSI, with the objective of defining open standards as to increase the competiveness of the enterprises into the concerned domain.

M441 "Smart Metering" mandate [2] has been running since March 2009 and will define open standards for water, gas and electricity meters communications. M468 on "Electric Vehicle Charging" [3] has been launched in 2010.gthers are under preparation with Experts Groups currently working on them; one of them concerning Internet of Things.

The ETSI approach of proposing a generic horizontal platformand applying it to vertical applications with complete stacks of protocols defined for many years, is not an easy task and compromises will need to be made. Nevertheless there are some trends that created from the work done by the standardisation committees:

- IPv6 is the target for transported datas
- Data models needs to converge towards a global "meta-model"
- Use of common and open APIs will help reducing the costs
- Internet Based Protocols and Web Appliances
- Tools developed in 3GPP will be of help when scaling to billions of Communicating Devices

- Low layers protocols will remain, as matching physical constraints and using different supports to carry the information (radio, cable, powerline, fiber) but they will converge on the upper layers, using IP convergence and Profiling.

12.1.4 Next Steps: The Way(s) Toward the Internet of Things

12.1.4.1 M2M service platform first release

There is always a series of compromises that conduct to the genesis of a commercial product, and standards obey the same rules. The set of standards defining the horizontal M2M platform will be delivered, taking into account time constraints and enabling technologies during the generation phase. The basic functionalities of the M2M system were implemented very quickly by the M2M committee, but then came the more complex ones that needed to intensify the liaisons with other SDOs to build on existing know-how and work carefully to include all the use-cases.

As the driving Domain was the domain of Energy, B2B (Business to Business) and B2B2C (Business to Business to Consumers) applications are key, with very high level of QoS (Quality of Service), including security, the first M2M release will implement, along the basic services capabilities, a set of advanced applications allowing among others, Remote Device Management, and Trusted Applications.

Anyway, as long as it is possible to define a minimum set of functionalities to be able to answer Energy Business Applications requirements, the first release will be "Energy compliant."

12.1.5 Roadmap Towards the Internet of things

The complexity of the M2M actors chain, see Figure 12.5, increased by the specificities generated by the main use cases (Smart metering, Smart Grid, Smart City, Connected consumer, e-Health, Automotive) has induced a slow-down of the standardization process.

Through the work already done and the stage of development reached, the functionalities that will be implemented further are listed below:

- Links with multiple Connectivity solutions
- Multi operator Value chain
- Identification/End to End Traceability

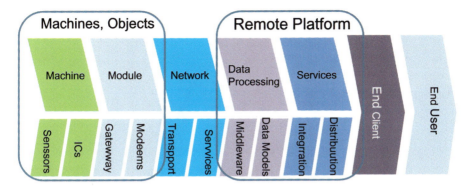

Fig. 12.5 M2M actors chain.

- Naming/Numbering/Adressing (full IP V6 is the target)
- Discovery Services
- Privacy constraints
- Interworking
- Scaling (billions of machines): Need of optimizing every constitutive block of the architecture
- Object Persistence into the network (even not connected).
- Total independence from the network and from the business applications
- APIs
- M2M language development

Improvements need to be made in each of these segments and with the help of a global multistage coordination, and with inputs and innovation from the collaborative R&D work and experimentations done in Europe, the pertinence of the approach will be validated.

Note: The author wants to thank ETSI for authorization of publication of elements included in published TRs, and Eurescom for having communicated the public hyperlinks towards P1856 [4] and P1957 [5] pre-standardisation projects related to M2M.

12.2 The "Internet of Things" Based on IPv6

The public IPv4 address space managed by IANA [6] has been completely depleted by Feb 1st, 2011. This creates in itself an interesting challenge when adding new things and enabling new services on the Internet. Without public IP addresses, the Internet of Things capabilities would be greatly reduced. Most discussions about IoT have been based on the illusionary assumption that the IP address space is an unlimited resource or it's even taken for granted that IP is like oxygen produced for free by nature. Hopefully, the next generation of Internet Protocol, also known as IPv6 will bring a solution.

In the early 90s, IPv6 was designed by the IETF IPng (Next Generation) Working Group and promoted by the IPv6 Forum since 1999. Expanding the IPv4 protocol suite with larger address space and defining new capabilities restoring end to end connectivity, and end to end services, several IETF working groups have worked on many deployment scenarios with transition models to interact with IPv4 infrastructure and services. They have also enhanced a combination of features that were not tightly designed or scalable in IPv4 like IP mobility, ad hoc services; etc catering the extreme scenario where IP becomes a commodity service enabling lowest cost networking deployment of large scale sensor networks, RFID, IP in the car, to any imaginable scenario where networking adds value to commodity.

With the exception of very few IPv6 experts, none of the previous discussions or research papers talked explicitly about the IPv4 address crunch and its impact on IoT or the open standards needed for its scalability, let alone ever mentioning IPv6 and its advanced IETF developments such as IPv6 adaptation layer over IEEE 802.15.4 (including header compression) known as 6LoW-PAN or IPv6 Routing Protocol for Low power and Lossy Networks (RPL) as the way forward. This paper wishes to restore some clarity in this area.

12.2.1 Open and Scalable Architectural Model

When embedding networking capabilities in "things", there are architectural decisions to be made that guarantee the "Internet of Things" is scalable, inclusive of several communication media, secure, future proof and viable for businesses and end-users. Several models can be discussed, (as reviewed below) but one clearly emerges as the best approach.

1. Closed or monolithic architectures

Today, when studying market segments already integrating networking capabilities in "things", one finds many ad-hoc alliances, proprietary, monolithic or closed protocols. Most of these focus only on the lower layers (Physical and Datalink) of the OSI model to transmit data or else define a complete network stack including application layers that only work on a single communication medium. Fortunately (or unfortunately), it was demonstrated long ago that the Earth is not flat, meaning in our context that a truly scalable network needs more than a single physical and data link layer to fit all of the needs and requirements of a diverse set of applications and use cases.

- A proprietary or monolithic protocol installation requires the upgrade of all networking components when moving to new technologies
- Markets remain fragmented with no interoperability unless done on the obvious protocol of choice — the Internet Protocol or IP.
- Mix & match of proprietary protocols are costly and inefficient, requiring protocol translation gateways for each protocol stack. Once again, it calls for translation to IP, but use of protocol translation gateways has been proven to be difficult to operate, manage and scale.

2. Why not a new protocol suite?

As the "Internet of Things" clearly represents a new generation of devices and applications, some people may think it would be better to begin from scratch and create a new architectural model. Although this may look attractive at first glance, such a proposal has a number of known issues:

- The "Internet of Things" is going to involve not only interactions between things, but also interaction between humans and their computers or their personal devices, as in the case with remote monitoring or remote control applications. By developing an "Internet of Things" separate from the actual Internet it will encourage market fragmentations as well as a "Balkanization" between both the existing Internet and "Internet of Things" in contrast to the convergence to IP that is now happening in areas such as telephony and TV.

In addition, open standards are key to the success of any protocol. It generally takes between 5 and 10 years from inception to production implementations. An entirely new architecture will delay the "Internet of Things" and be an obstacle to rapid market adoption.

3. The Internet Protocol (IP)

Although certainly not 100% perfect, expecting researchers and others to enhance the protocol suite, the TCP/IP model has demonstrated:

- Capacity to be deployed on a very large scale, aka "The Internet"
- Centralized (i.e.,: An Intranet) or distributed (i.e.,: The Internet) deployment models
- Versatility to handle all types of traffic, including critical traffic such as voice and video.
- Extensive interoperability as IP runs over most if not all available industry standard network links — wireless (802.15.4/6LoWPAN, Wi-Fi, 3G, WiMax,...) and wired (Ethernet, Sonet/SDH, serial,...).
- Open process of standardizations through the IETF and associated standard bodies, enabling consensus on enhancements and interoperability.
- Future proofing through the adoption of a next generation of IP protocol, aka IPv6.
- Established application level data models and services which are well understood by software developers and widely known to the public through worldwide web applications. The diversity of applications with the web services paradigm is a prominent one thanks to its authorisation of distributed computing across platforms, operating systems, programming languages, and of course vendors and products.
- Established network services and architectures for higher-level services and scalable deployments:
    - Naming, addressing, translation, routing, services discovery, load balancing, caching, and mobility.
    - Diversity of well understood security mechanisms at different layers and different scopes.
    - Diversity of network management tools and applications.

12.2.2 Internet Protocol Version

Once it is agreed that IP is the appropriate architectural model, comes the question of recommending an IP version. The Internet today is going through a transition due to the IPv4 address space exhaustion [OECD, IPv4add-report]. With little existing legacy in the "Internet of Things", there is an opportunity to promote IP version 6 (IPv6) as the de-facto IP version for the "Internet of Things." This recommendation would be fully aligned, such as:

- U.S. OMB [7] and FAR [8]
- European Commission IPv6 recommendations [9]
- Regional Internet Registry recommendations [10]
- IPv4 address depletion countdown [11]

Adoption of IPv6 for the "Internet of Things" benefits from:

- A huge address space accommodating any expected multi-millions of deployed "things."
- "Plug & Play" capabilities as IPv6 protocol suite enhance provisioning mechanisms suitable to "Things." Flexibility of address configuration as demonstrated by

    o Stateless IPv6 address configuration

    o DHCP Individual address configuration

    o DHCPv6 Prefix Delegation + Stateless IPv6 configuration

- IPv6 is the future IP addressing standard: De facto IP version supported by new physical and data link layers such as IEEE 802.15.4 and IEEE P1901.2 through the standardized IP adaptation layer — IETF 6LoWPAN WG — which only defines IPv6 as protocol version. No IPv4 standard equivalent has been specified.
- De facto IP version for the standardized IETF Routing Protocol for Low Power and Lossy Networks (RPL) — IETF RoLL WG — as it is an IPv6-only protocol.
- If needed, IPv6 over IPv4 tunnelling or IPv6-IPv4 translation could be achieved when "things" would need to communicate with legacy back-end systems not supporting IPv6. It should be noted that all recent operating systems do support the new IP protocol version [OS], making the "translation" requirement as the least preferable solution.

However, an IPv6 address is not a unique identifier and should not be considered to identify a "thing" as the address is expected to change when that "thing" gets connected in a different location or network.

12.2.3 Security and Privacy

By adopting the TCP/IP architecture for the "Internet of Things", all of the lessons from the years spent securing private and public IP infrastructures will apply to the new environments. However, some may consider the "Internet" concept as unsecured or lacking privacy. For this reason, it is important to remind people that security and privacy are a multi-faceted challenge.

- Security of "things"
  As for any device attached to a network, security represents a multi-layered challenge to get addressed by owners or network managers. It ranges from

    o Securing the physical access to a "thing."

    o Authenticating the data link, network and application access.

    o Encrypting data on data links and network links when necessary — in accordance with regulations.

  All existing mechanisms today require appropriate documentation and enhancements for true plug & play.
- Network security
  Connecting "things" to an IP environment means that all network design and policy already defined on the intranet or when providing access to the Internet will apply to the additional sub-networks hosting those "things." It means that from day one, authentication, access control, firewall and intrusion detection mechanisms should fully be operational for the "Internet of Things."
- Privacy
  The "Internet of Things" is no more than an additional layer of devices connecting to the Internet. The fact that "things" implement an IP stack does not mean they have to be fully reachable over the Internet. As for any node, it is the responsibility of the owner or

network manager to decide if the "things" fully participate in the Internet or stays isolated on an intranet.

In addition, when "things" are fully reachable over the Internet, it is still important to decide who can communicate with them, including new mechanisms where "things" could share a physical network, but be managed and accessed by dedicated entities. Once again, similar mechanisms and policies already set-up for intranet and Internet accesses apply. However, additional standards may be required to see new business models and usage benefiting from the "Internet of Things." For example, imagine smart meters for electricity, gas and water sharing the same end-user's broadband connection. It certainly does not mean that all entities want to share data related to their business between each other. At the same time, they may want to allow access by the "end-user's things" to some of the information captured by the smart meters allowing local actions to be taken for energy savings. This would require new standards efforts to create such models. For example Zigbee/IP selected an IPv6 (6LoWPAN/RPL) stack for its Smart Energy Profile 2.0

12.2.4 Applications & Services

A successful adoption of the "Internet of Things" will be largely dependent of the availability of applications and services. Similarly to the actual Internet, traffic flows are expected to range from:

- "Things" to back-end servers
- "Things" to end-user's browsers
- "Things" to "things"

Ease of use is a key criteria to address the mass market. It cannot be expected that an average end-user would manually enter IP(v6) addresses when installing new "things" on a network. Likewise, it may not be assumed that "things" can permanently store all addresses of other "things" in a network. For those reasons, it is believed there is a strong need to develop and standardize naming services for the "Internet of Things." However, considering the range of traffic flow, naming services must accept address/name resolution for all kind of communications as previously listed. It cannot be envisaged that an "Internet of Things" naming services would be established disconnected from the existing DNS.

Today, the success of the Internet is largely due to the adoption of worldwide web tools, the new generation of collaborative tools or web 2.0 now representing the next step. It is expected that standard web services will be widely implemented by the first generation of applications on the "Internet of Things", especially for management, data presentation and analysis.

Some applications may not be web-oriented per se, but would enable communications between two "things", satisfied with a given definition at the application layer.

In some application domains, do exist applications layers that describe device profiles and capabilities in an optimized fashion for low-resource environments. Often times, they are specific to certain devices and interaction types (control-oriented apps like lighting and temperature setting). The IETF CoRE WG and Extended XML (EXI) are examples of enabling such highly optimized and very descriptive application layers to run on any IP device, including those that are resource-constrained — where 6LoWPAN may be used to provide the underlying IP connectivity — as well as those that have more traditional resource footprints like computers, handhelds, and servers.

12.3 Paving the Way to Smart IPv6 Buildings

By 2020, the Internet is expected to connect 50 to 100 Billion smart things and objects [12], with an increasing volume of circulating data and information. The current evolution is heading towards an increasingly interconnected, mobile, pervasive and ubiquitous Internet of Things, with potentially billions of heterogeneous things and devices communicating with each other and working together. Yet, the current Internet Protocol (IPv4), is limited to 4 Billion IP public addresses,[1] — and by the beginning of 2011, IANA will have distributed its last IPv4 addresses blocks to the regional organizations (RIR) [13]. They will be fully attributed to ISPs and other organizations by the beginning of 2012 according to the Number Resource Organization (NRO) [14]. The Internet Protocol version 6 (IPv6) being the only available alternative to satisfy the growing needs for IP addresses, many countries are preparing a large scale transition towards a dual-stack IPv4/IPv6 network infrastructure.[2] The

---

[1] IPv4 relies on 32 bits addresses; IPv6 provides an address space of 128 bits.
[2] Countries like China, Japan, Korea have developed large scale IPv6 network. The US administration requires from all their suppliers to provide IPv6 compliant ICT equipment.

IPv6 provides enough addresses (2^{128} IPv6 addresses) to provide each and every device and "smart thing" on earth with its own public IP address.

Beyond the scalability issue and the possibility to get rid of Network Address Translation (NAT) barriers, the IPv6 carries new features, such as look-up, self-configuration, security and authentication mechanisms, which could make the deployment of Internet of Things easier and more reliable. The IPv6 enables end-to-end communication, in which any IPv6 "smart thing" can connect to any other IPv6 device or system from any place and at any time. It enables the possible extension of the Internet to any device, sensor or actuator, which can become real nodes of the future Internet. In other words, the IPv6 (together with its lighter version 6LowPAN) will make possible to move from a network of servers, to a global network of smart things.

12.3.1 The International Cooperation House Project and the Smart IPv6 Building Approach

Mandat International [15] is a foundation based in Geneva, which aims at promoting international cooperation and supporting the participation of delegates in international conferences. The foundation also develops it own research activities in areas such as ICT and sustainable development, with a strong interest for the IPv6.

Every year, about 200.000 delegates and experts come to Geneva to attend international conferences. In 2005, Mandat International started to work on a project of new building to accommodate and support delegates from developing countries attending UN conferences: The International Cooperation House project [16]. This 11 floor building will be located in the heart of the international organizations area of Geneva, close to the UN, with:

- Budget accommodation for experts and delegates coming from developing countries.
- Office space for organisation working with the UN system.
- Public infrastructure: Conference room, exhibition area, restaurant, etc.

The project will benefit from a high visibility and will mix three kinds of publics: Delegates, international civil servants and the local population. It appears that this project will constitute an ideal opportunity to promote clean technologies and to explore the potential of the IPv6 for making the building smarter and more energy efficient. The foundation accepted to turn it into a platform for research cooperation in order to research and showcase innovative forms of interactions demonstrating the full potential of the IPv6.

Very quickly, a concept of "Smart IPv6 Building" emerged, in which the IPv6 serves as an integrator between sensor networks, actuators and building automation devices,- as well as with other components such as on-line information, web services and user interfaces. It takes into account the progresses in nanotechnology supporting the evolution towards more and more pervasive intelligence to be deployed. In the Smart IPv6 Building concept, the IPv6 is used as a common ground and as an enabler to address various objectives, such as:

- Monitoring the activity and energy consumption in the building.
- Reducing energy consumption through a smarter energy management.
- Providing an interactive environment with distributed information and decentralized communication among objects.
- Enabling innovative interactions and networking among delegates, as well as with the building environment.
- Managing access and security.
- Showcasing new technologies related to the Internet of things

The Smart IPv6 Building approach paves the way to a more integrated Internet of things in homes, with potential benefits in terms of energy saving, comfort, and security. This vision is currently being implemented by Mandat International together with interested academic and industrial partners, and with the active support of the IPv6 Forum.

12.3.2 The Universal Device Gateway Development

Buildings are traditionally heterogeneous and fragmented environments, gathering all sorts of smart things. They are complex environments with all kinds of systems that coexist while using incompatible communication protocols. This makes their management complicated and strongly limits the possible interactions. To address this interoperability challenge, a Swiss national research project has been launched with a consortium gathering Mandat International, the University for Applied Sciences Western Switzerland, Archimède Solutions and Smart Home. The project has successfully developed a Universal Device Gateway (UDG) [17] which uses IPv6 to integrate and to interconnect heterogeneous subsystems using different communication protocols and standards into a shared semantic framework. The research project allowed the integration through IPv6 of about ten different communication protocols that are common in buildings. The UDG demonstrated new forms of cross-domain interactions, enabling heterogeneous devices such as mobile phones, RFID tags, ZigBee sensors, KNX actuators, X10 switches and UPnP media components to interact with each others. The UDG contributes to the transformation of buildings into smarter, more integrated and "user-friendly" environments. It enables new forms of cross-domain interactions and simplifies theirs management.

12.3.3 Hobnet Research Project

The FP7 European research project Hobnet has been launched with the University of Patras, the University of Geneva, the University of Dublin, the University of Edinburgh, Ericsson, Sensinode and Mandat International. Hobnet is addressing several research challenges related to the integration of heterogeneous sensor networks into an IPv6 environment. It is addressing the specific needs and requirements of the future International Cooperation House project and will exploit the FIRE infrastructure. The project's research addresses algorithmic, networking and application development aspects of Future Internet systems of tiny embedded devices, including:

- An all IPv6/6LoWPAN infrastructure of buildings and how IPv6 can integrate heterogeneous technology (sensors, actuators, mobile devices etc);
- 6lowApp and its standardization towards a new embedded application protocol for building automation;
- Novel algorithmic models and scalable solutions for energy efficiency and radiation-awareness, data dissemination, localization and mobility;
- Rapid development and integration of building management applications;

- Support for the deployment and monitoring of resulting applications on FIRE test beds.

12.3.4 Setting Up a Smart IPv6 Building Testbed

In order to explore and validate the Smart IPv6 Building approach, Mandat International has decided to provide one of its current welcome centers as a living test-bed. It is a 4 floor building accommodating delegates from developing countries. The building comprises various functional areas, including accommodation, office space, work infrastructure, a meeting room, a kitchen and other facilities. It is exposed to very similar end-users to the ones planned in the future International Cooperation House. Beyond the technical tests and validations, it gives the opportunity to address the multi-cultural end-user acceptance with delegates from all around the World. The building is being currently IPv6 empowered and equipped with sensors, which will serve to test different use case scenarios and to evaluate their impact, including in terms of energy saving. The test-bed will integrate various equipments and devices from academic and industrial partners, and will be used for the validation of the Hobnet research project. The FP7 project 6DEPLOY-2 is also supporting this deployment with training and practical assistance for the network configuration. 6DEPLOY-2 will also promote this use case on its web site as an example of an IPv6 deployment within Smart Buildings.

12.3.5 The Way Forward

The Smart IPv6 Building concept gathers like minded partners interested to research and develop this innovative approach. It is an opportunity to test innovative paradigms, solutions and products in an environment with real end-users, and to promote those solutions internationally. It also enables to identify and address new research challenges through collaborative research projects. More information is available on [18].

12.4 Validation and Interoperability Challenges for IoT

Non-interoperability [19] impedes the sharing of data and the sharing of (computing) resources, causing organizations to spend much more than necessary on data, software, and hardware. As organizations today are under "economic constraints," the issue of non-interoperability is one that obviously needs to be resolved quickly.

Organizations seek to avoid unnecessary risks. Non-interoperability increases technology risks, which is a function of:

- The probability that a technology will not deliver its expected benefit and,
- The consequence to the system (and users) of the technology not delivering that benefit.

Risk assessment must take into account evolving requirements and support costs. Some technology risks derive from being locked into one vendor, others from choosing a standard that the market later abandons. The direct risks associated with non-interoperability are real-world risks. Today, lives and property depend upon digital information flowing smoothly from one information system to another. No single organization produces all the data (so it is inconsistent) and no single vendor provides all the systems (so the systems use different system architectures, which are usually based on different proprietary interfaces). Thus, there is the potential for real world disorder.

12.4.1 Interoperability

What is important for any communication technology is even more crucial for IoT as it will have massive deployment and communication, and will be based upon:

1. Several access technologies some time "radio" ones which would lead to potential radio co-existence issues,
2. Several protocols some are just emerging and not "validated" and
3. New architectures, data structure, naming convention, governance, etc all new features or concept needed to be validated even before being ready for final "interoperability checks."

It should be noted that pragmatic initiatives to validate solutions and even standards, to report on implementation and market driven issues or to just improve interoperability are on the organisation of interoperability events.

But to profit fully from an interoperability event, some "validation" activities in using dedicated methodologies would be needed as experience shows

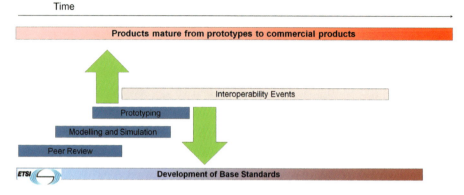

Fig. 12.6 Standards development and the product life cycle.

that interoperability cannot be ensured properly without *unambiguous specification, market accepted unambiguous test specifications* and finally some pragmatic evidence of *multi-vendors interoperability*.

More detail can be found in the ETSI White Paper on Interoperability [20] introducing that in the last few years the nature of using technologies based on stable market accepted standards has fundamentally shifted. Complex technologies are implemented from 'islands of standards', sometimes coming from many different organisations, sometimes comprising hundreds of different documents.

Resulting products need to fit into this technology standards map both on the *vertical* and the *horizontal* plane. Applications need to interoperate across different domains, as well as directly interworking with underlying layers. This complex ecosystem means that issues of interoperability are more likely to arise and the problem is more critical than ever.

The White Paper will document the need to take care of the consistency of the base "specifications" or standards and on the importance of unambiguous and well accepted test specifications. This part is often under-estimated or left alone to the market forces and in such cases lead to large technologies deployment failures. Although the market forces are pushing for cost-effective developments and these concern huge collective investments on base specifications, test specifications, test tools, validation and certification programmes, the total cost of "doing nothing" is somewhat larger than starting from the beginning. *That also requires research efforts in line with the complexity of*

Fig. 12.7 RFID Interoperability event Beijing April 2009.

the research topics (e.g., FP7 Walter on UWB testing on its complexity supporting the technology and its deployment).

Finally, the use of interoperability events in the standardisation process has proved to be an exceptionally useful and popular way to practically and efficiently verify/validate standards and equipment to improve the chances of interoperability. They provide direct feedback to both the standards developers and the product developers in a way that no theoretical (paper) activity can.

In the past 10 years or so, some organisations like ETSI (www.etsi.org/plugtests +150 events organised) but also industrial fora have organized and ran interoperability events. Much positive feedback has been gathered on the benefit of this pragmatic approach to improve the interoperability of products and services, the quality of the specifications and the whole standard-making process in supporting successful deployment of the technologies researched by the FP projects.

12.5 Conclusions

The work done in ETSI M2M proposes a common architecture for the M2M service platform, with three interfaces and their corresponding

APIs: The interfaces to the M-devices, to the Transport layer and to the Business application layer. The main objective is to ensure interoperability of the service platform with all types of devices, whenever they are directly connected to the Network (D-Devices) or through a gateway (D'-Devices), and whenever the underlying access and transport network is. On the opposite side, the platform will be able to read different data models using a common methodology and dedicated APIs. This convergent approach will integrate in the next future Tags and Readers System layer, as to define one common service layer for Machines, Devices and Tags.

The definition and adoption of standards-compliant IPv6 stack and standards-compliant Web Services are the key enablers for a "thing" to be effectively operated and managed in the broader "Internet of Things."

Issues on Interoperability and IoT validation must not be underestimated and appropriate research in these domains must be undertaken as technologies become more complex, aswell as testing and interoperability required innovative approaches too.

References

[1] ETSI portal, http://portal.etsi.org/portal/server.pt/community/M2M/319.
[2] EC Mandates M/441, http://www.etsi.org/WebSite/AboutETSI/RoleinEurope/ECMandates.aspx
[3] EC Mandates M/468.
[4] Eurescom project, http://www.eurescom.eu/Public/Projects/P1800series/P1856/default.asp
[5] Eurescom project, http://www.eurescom.eu/Public/Projects/P1900series/P1957/default.asp (Project submitted in 2010).
[6] The Internet Assigned Numbers Authority (IANA), http://www.iana.org
[7] Executive Office of the President, Office of Management and Budget, "Memorandum for the chief information officers", Washington D.C., August 2, 2005, http://www.whitehouse.gov/sites/default/files/omb/assets/omb/memoranda/fy2005/m05-22.pdf
[8] Department of Defense, General Services Administration, National Aeronautics and Space Administration, "Federal Acquisition Regulation; FAR Case 2005-041, Internet Protocol Version 6 (IPv6)", Volume 74, Nmb 236, pp 65605-65607, December 10, 2009, http://edocket.access.gpo.gov/2009/E9-28931.htm
[9] European Commission IPv6 recommendations, http://ec.europa.eu/information_society/policy/ipv6/index_en.htm
[10] American Registry for Internet Numbers (ARIN), "IPv4/IPv5: The Bottom Line", https://www.arin.net/knowledge/v4-v6.html
[11] The IPv4 Depletion Site, http://www.ipv4depletion.com/?page_id=147

[12] Vision and Challenges for Realising the Internet of Things, March 2010, European cluster CERP-IoT, p13, and Ericsson perspective for the IoT, http://gigaom.com/2010/04/14/ericsson-sees-the-internet-of-things-by-2020/ and http://www.ericsson.com/news/1403231.
[13] IPv4 Address Report, http://www.potaroo.net/tools/ipv4/.
[14] Number Resource Organisation (NRO), "Remaining IPv4 Address Space Drops Below 5%", 18 October 2010, http://www.nro.net/news/remaining-ipv4-address-space-drops-below-5
[15] Mandat International, www.mandint.org
[16] International Cooperation House, www.internationalcooperationhouse.org
[17] Universal Device Gateway, www.devicegateway.com
[18] Smart IPv6 Building, www.smartipv6building.org
[19] The Havoc of Non-Interoperability An Open GIS Consortium (OGC) White Paper Mark Reichardt — Executive Director
[20] Achieving Technical Interoperability — The ETSI approach' ETSI White Papers

Index

6LoWPAN, 298, 300, 301, 303–305, 307

architectural, 298–300
architecture, 258, 261, 267, 269–272, 274, 275
architecture model, 144, 175, 176, 201
augmented reality, 188

cloud computing, 18–20, 28, 34, 35, 42
communications, 13, 18, 20–23, 29, 40

discovery and search engines, 152, 222

embedded systems, 18, 20, 21, 23, 37, 46
ethics, 83, 84, 98
European research cluster, 2, 4, 6

future internet, 10, 20, 32, 35

governance, 65, 66, 74, 86, 87, 89, 143, 171, 257, 258, 265, 270

Internet, 287, 289, 295–307, 312
Internet of Things, 1, 2, 4, 10, 12–18, 21, 22, 25, 27–32, 34, 35, 37–43, 50
Internet of Things conceptual framework, 10
Internet of Things technologies, 231, 234, 236, 237
interoperability, 257, 258, 263, 265, 266, 287, 288, 299, 300, 306, 308–312
IoT, 109, 112, 113
IoT activity chains, 3
IoT architecture, 3

IoT European research projects, 57
IoT governance, 3
IoT household appliances, 184
IoT knowledge integration, 2
IPv6, 287, 295, 297, 298, 300–308, 312

law, 77–80, 88, 90, 93

machine to machine communication, 288
middleware, 258, 260–262, 265, 269, 273–275, 282
mobile devices, 179

nanoelectronics, 18, 20, 23
network abstraction, 249, 250
network and communication, 147
network technology, 215
network virtualization, 241, 242, 244, 246–255
NFC, 181–184, 192

objects search, 190

performance, 245, 246, 252
privacy, 53, 57, 61, 65, 67, 74, 76–81, 83–86, 90–92, 98
programmability, 243, 248
protocol, 295, 296, 298–301, 304, 306, 307, 309

re-configurability, 251

scalability, 245, 246, 247
scalable, 293, 298–300, 307
science, 109
security, 246, 248–250

security and privacy, 156, 226
self regulation, 90, 104
sensing, 179, 186
sensors, 11, 15, 18, 20, 21, 23, 28, 29, 34–38, 45
service and organizational layer, 109
smart healthcare, 189, 190
smart phones, 11, 18, 20
smart things, 106
socio-economic ecosystems, 171, 281
software Platform for IoT, 273
software technologies, 18, 20
standardisation, 288, 295, 297, 311
standards, 55, 68, 72, 74, 78, 81, 82, 84, 86, 89, 90
strategic research agenda, 17, 50

technicism, 109
topology awareness, 249, 251

traceability management, 115, 136, 137

ubiquitous communicator, 118, 119, 122, 123, 127, 128
ubiquitous computing, 115–117, 122, 124, 125, 138
ubiquitous ID architecture, 115, 117–121, 124
ucode, 117–120
ucode information server, 118–120, 123
ucode resolution server, 118–120, 123, 124
ucode tag, 118, 121, 122, 130, 137

validation, 287, 293, 308–310, 312

RIVER PUBLISHERS SERIES IN COMMUNICATIONS

Other books in this series:

Advances in Next Generation Services and Service Architectures
ISBN: 978-87-92329-55-4

Advanced Networks, Algorithms and Modeling for Earthquake Prediction
ISBN: 978-87-92329-57-8

Telecommunications in Disaster Areas
ISBN: 978-87-92329-48-6

Single and Cross Layer MIMO Techniques for IMT-Advanced
ISBN: 978-87-92329-50-9

Multihop Mobile Wireless Networks
ISBN: 978-87-92329-44-8

Adaptive PHY-MAC Design for Broadband Wireless Systems
ISBN: 978-87-92329-08-0

Towards Green ICT
ISBN: 978-87-92329-34-9

Planning and Optimisation of 3G and 4G Wireless Networks
ISBN: 978-87-92329-24-0

Link Adaptation for Relay-Based Cellular Networks
ISBN: 978-87-92329-30-1

Principles of Communications: A First Course in Communications
ISBN: 978-87-92329-10-3

*Single- and Multi-Carrier MIMO Transmission for
Broadband Wireless Systems*
ISBN: 978-87-92329-06-6

*Ultra Wideband Demystified: Technologies, Applications, and
System Design Considerations*
ISBN: 978-87-92329-14-1

Aerospace Technologies and Applications for Dual Use
ISBN: 978-87-92329-04-2

4G Mobile & Wireless Communications Technologies
ISBN: 978-87-92329-02-8

Advances in Broadband Communication and Networks
ISBN: 978-87-92329-00-4

CPSIA information can be obtained
at www.ICGtesting.com
Printed in the USA
262542LV00001B